Navigating Health Challenges in Space

Raimondo G. Russo

Navigating Health Challenges in Space

Systems Adaptations, Alterations and Medical Care

Springer

Raimondo G. Russo
Residenza Campo
BASIGLIO, Italy

ISBN 978-3-032-03619-3 ISBN 978-3-032-03620-9 (eBook)
https://doi.org/10.1007/978-3-032-03620-9

This Springer imprint is published by the registered company Springer Nature Switzerland AG
The registered company address is: Gewerbestrasse 11, 6330 Cham, Switzerland

L'amor che move il sole e l'altre stelle
Dante Alighieri (*)
Love that moves the sun and the other stars

(*) Paradiso, XXXIII, v. 145.
(Last verse of Paradise in the Divine Comedy, 1304–1306)

To Ma and Pa

Foreword

The topic sometimes referred to as "phanta-medicine" is rapidly becoming reality, as not only are astronauts exploring Space, but efforts are even being made to turn Space into a target for adventure travel. This book, "Medicine in Space" by Raimondo G. Russo, is therefore timely. Space medicine, or aerospace medicine as it is sometimes called, is not really a new medical speciality, but it involves a new "class" of medicine dealing with the situations encountered by man living in Space: right now there are at least three orders of these.

The first is the speed of travel in Space. Eight kilometers per second is about 29 times an aircraft's speed, which may reach 0.27 km/sec. The acceleration required to reach and maintain this may be dangerous to various body organs, especially considering it is constant, with no pauses unless the vehicle lands on a planet.

The second is microgravity, which puts spaceflight passengers under a new type of stress related to how they can move, in ways unusual for life on Earth. There is also a scarcity of oxygen but high levels of carbon dioxide, which may affect cerebral functions in particular. Then there may be alterations to the circadian rhythm, causing sleep problems, bearing in mind that at least seven hours of sleep is important to clear the brain of the toxic products of its metabolism. The third problem for mankind is space radiation, which can vary at different extraterrestrial altitudes.

These new questions call for wholesale revision of medicine to date, starting from prevention. It will be extremely important to establish what preparation is needed.

What kind of alimentation is best, not only to prevent problems, before and during travel in Space, when the consumption of calories changes considerably.

Questions of diagnosis and intervention will be quite different during space travel: it would certainly be hard to stop the vehicle and take appropriate measures, but ample information is available via telemedicine. Obviously, all the equipment likely to be needed must be kept on board.

Another point is what kind of drugs should be available for space travelers? How much do we know about the pharmacokinetics, efficacy, and toxicity of drugs when used in Space? When many people travel in Space, will they suffer the same diseases as on Earth? And after traveling in Space, will it be easy to return to what was "normal life," or will they need a period of adaptation?

We certainly cannot stop our curiosity and interest in Space, but the health needs for travelers are important and numerous. Clearly, questions are not going to be answered by National Health Systems. At present problems encountered by Space passengers should be treated privately, certainly not backed by public money.

President of Mario Negri Institute Silvio Garattini
Milan, Italy
27 October 2024

Acknowledgments Here I want to thank Prof. Silvio Garattini who, with his comments in the Foreword of this book, truly supported the relevance and also the modernity of the text.

Moreover, I also want to point out the great help offered to me by Dr. Judith Baggott for corrections and for the best expression in correct English.

Contents

Part I

The Dawn of the Space Age

Introduction

1

[1] Space medicine or aerospace medicine is a branch of medicine that deals with the possibilities of life for man in extraterrestrial space and the means available to ensure survival in environmental conditions that are hostile or fundamentally different from those on Earth.

Mankind's interest and study of extraterrestrial space throughout history dates back to remote times, which are hard to identify. Since the first testimonies relating to prehistoric times, man has observed and represented the celestial elements, but only from the second half of the twentieth century, with the evolution of aerospace technologies, has it been possible to physically begin space exploration.

We began to talk about space medicine when the development and increasing power of various means of propulsion were initially only envisaged, but it soon became possible for us to take ourselves into interplanetary spaces.

The fundamental problems for protecting life in space are mainly due to the lack of the Earth's atmosphere and therefore of the oxygen essential for life and food. There is also the lack of gravity which can lead to psychological and physiological problems. A satellite or spacecraft must reach a speed of at least 8 km/s to enter orbit around our planet, and this requires an acceleration that must not harm an astronaut.

Aerospace medicine has been a recognized specialty since 1955. Space medicine focuses specifically on preventive medical care of astronauts and spaceflight participants and performance optimization. The spaceflight environment poses many unique stresses for the human body, including G forces, microgravity, unusual atmospheres such as low pressure or high carbon dioxide, and space radiation.

Space medicine applies space physiology, preventive medicine, primary care, emergency medicine, acute care medicine, austere medicine, public health, and toxicology to prevent and treat medical problems in space.

[2] Space and particularly long-duration space travel, such as future missions to the Moon or Mars, pose an enormous challenge for human health and will require major innovations developed in medicine. From prosthetics to telemedicine, many

R. G. Russo, *Navigating Health Challenges in Space*,
https://doi.org/10.1007/978-3-032-03620-9_1

innovations developed during NASA space missions have already found application in the medical field. Even short-term spaceflight in low Earth orbit induces changes in the human organism similar to those caused by a longer stay. Among these are the high levels of cytokines (molecules that modify the activity of cells in response to a stimulus) and changes in the expression of genes linked to activation of the immune system and the repair of DNA damage. More than 95% of these parameters return to normal once back on Earth, though some remain altered for longer—at least 3 months after the mission.

Space innovation draws on the fields of medicine and life sciences, with spectacular results. The advanced technologies developed in space exploration, specially by the US space agency NASA, were originally designed for the missions of its own astronauts, but have since been applied commercially and have been "revolutionizing" the medical sector for decades.

What are referred to as "spinoffs" of this progress constitute a series of scientific and technological discoveries and applications that offer unprecedented opportunities to improve human health and save lives. From precision medical imaging to cutting-edge prosthetics, from telemedicine to early disease diagnosis, space mission spinoffs are changing the face of modern medicine. In brief, NASA spinoffs represent exceptional innovation, creating synergy between space exploration and medical progress. These advanced technologies are transforming medicine, improving diagnoses, treatments, and healthcare in general: exploiting their potential is a promising path towards the future.

Aerospace medicine is a branch of medicine born in the 1950s with the first launches of living beings into space. Man reached geocentric orbit on April 12, 1961, with the flight of Soviet cosmonaut Yuri Gagarin. Since then, human missions have become increasingly frequent and long-lasting, even reaching the lunar surface in 1969. Since the 1970s, longer flights have been carried out and space stations have been developed in which astronauts remain in orbit for weeks or months.

Science proceeds diligently in the search for new treatments for known pathologies, defining and clarifying those that are still not clear and launching new and original investigations to detect any that are still unknown, and deal with them. The findings quickly became established worldwide.

Meanwhile, further predictive studies for the identification and determination of diseases in space increasingly move forward. The main point is to establish what it is all really about. This implies formulating a definition which, at least initially, must refer to something that is still unknown and therefore indefinable. However, there are already signs—by no means negligible—that the first steps are being taken.

Living and working in space entails a few problems: weightlessness, radiations [3–5] and solar radiation, difficulties in carrying out physiological needs and taking care of personal hygiene, and nourishment are some examples of settings in which aerospace medicine can be called upon.

The main priorities of a "journey" into space are the search for extraterrestrial life, the colonization (and exploitation) of celestial bodies, and the development of new technologies which will find a wider field of application in other sectors as well. Among the latter, the biomedical sector obviously has a major role.

Why Do We Go To Space ?

At the US *National Aeronautics and Space Administration (NASA)*, we can explore the secrets of the universe for the benefit of all, creating new opportunities and inspiring the world through discovery. NASA's exploration vision is anchored in providing value for humanity by answering some fundamental questions: Why are we here? How did it all begin? Are we all alone? What comes next? Then, almost as an addendum: How can we make our lives better?

NASA was created more than half a century ago to begin answering some of these questions. Since then, space exploration has been one of the most unifying, borderless human endeavors to date. An international partnership of five space agencies from 15 countries operates the International Space Station (ISS), and two dozen countries have signed the Artemis Accords [6] signaling their commitment to shared values for long-term human exploration and research at the Moon.

> Through space exploration, we gain a new perspective to study Earth and the solar system. We advance new technologies that improve our daily lives, and we inspire a new generation of artists, thinkers, tinkerers, engineers, and scientists. [7]

Exploring the Unknown

The seven-volume documentary history listed here [8–14] is an essential reference for anyone interested in the history of the US civil space program and its progress over time. Each volume deals with specific issues in the development of the space program and includes more than 110 key documents, many of which were published for the first time. Each is introduced by a headnote providing context, bibliographical details, and background information necessary to understand the document. These are organized into major sections, each beginning with an introductory essay that keys the documents to major events in the history of space exploration. International cooperation on Artemis is intended not only to bolster space exploration but also to enhance peaceful relationships between nations. Therefore, at the core of the Artemis Accords is the requirement that all activities will be conducted for peaceful purposes, in accordance with the Outer Space Treaty.

The Mostly Unknown Universe

Around 95% of the universe is still a mystery with *unknown exotic particles and* new forces awaiting discovery. The observable universe is a spherical region of space consisting of all matter that can be known from Earth or its space-based telescopes and exploratory probes at the present time. Initially, it was estimated that there may be two trillion galaxies.

According to calculations, the current distance to particles emitting the cosmic microwave background radiation (CMBR) which represents the radius of the visible universe, is about 14.0 billion parsecs—around 45.7 billion light-years. We do not know the true size of the universe, and it may be infinite. Some parts of the universe are too far away for the light emitted since the Big Bang to have had enough time to reach Earth or space-based instruments and therefore lie outside the observable universe [15].

The biggest mystery in space is that the largest amount of matter is dark and consists of unknown particles. If that was not mysterious enough, the vacuum of empty space is filled with a mysterious dark energy that accelerates the expansion of the universe. (Jan. 25, 2024)

Bibliography

1. Wikipedia, the free encyclopedia. Medicine in Space.
2. Shirah B, Bukhari H, Pandya S, et al. Benefits of Space Medicine Research for Healthcare on Earth. Cureus. 2023;15(5)
3. Chancellor JC, Scott GB, Sutton JP. Space radiation: the number one risk to astronaut health beyond low earth orbit. Life (Basel, CH). 2014;4(3):491–510.
4. Guidance on radiation received in space activities: recommendations of the National Council on Radiation Protection and Measurements. Space Radiation Report. 1998.
5. Wikipedia. Cosmic ray; NASA Why Space radiation Matters, Apr 13, 2017; Radiation and your Health. Facts about radiation from Space.
6. The amount of radiation astronauts are subjected to in space varies between the dose a patient receives during an x-ray and a single radiation therapy session: from 1 to 100 milligrays. (May 9, 2023). The intense amount of radiation a they are exposed to: According to NASA, that's the equivalent of between 150 and 6,000 chest X-rays.
7. NASA, Ed. Shea G, Why go to Space, Sep 22, 2023.
8. Selected documents in the history of the U.S. Civil Space Program volume i: organizing for exploration. John M. Logsdon, Editor with Linda J. Lear, Jannelle Warren-Findley, Ray A. Williamson, and Dwayne A. Day.
9. Selected documents in the history of the US. Civilian Space Program volume II: external relationships. John M. Logsdon, Editor with Dwayne A. Day and Roger D. Launius.
10. Selected documents in the history of the U.S. Civil Space Program volume III: using space. Edited by John M. Logsdon, with Roger D. Launius, David H. Onkst, and Stephen J. Garber. 1998. PDF format (3 MB).
11. Selected documents in the history of the U.S. Civil Space Program volume IV: accessing space. Edited by John M. Logsdon, with Ray A. Williamson, Roger D. Launius, Russell J. Acker, Stephen J. Garber, and Jonathan L. Friedman. 1999. PDF format (4.2 MB).
12. Selected documents in the history of the U.S. Civil Space Program volume V: exploring the cosmos. Edited by John M. Logsdon, with Amy Paige Snyder, Roger D. Launius, Stephen J. Garber, and Regan Anne Newport. 2001.
13. Selected documents in the history of the U.S. Civil Space Program volume VI: space and earth science. Edited by John M. Logsdon, with Stephen J. Garber, Roger D. Launius, and Ray A. Williamson. 2004.
14. Selected documents in the history of the U.S. Civil Space Program volume VII: human spaceflight. Edited by John M. Logsdon, with Roger D. Launius. 2008.
15. Bars I, Terning J. Extra dimensions in space and time (Multiversal journeys). Springer; 2010.

2 Brief History of the Conquest of Space (and of Spatial Medicine)

Rockets have interested scientists for centuries. The Chinese used them as weapons as early as the eleventh century. Russian scientist Konstantin Tsiolkovsky (1857–1935) [1] theorized in 1880 that rockets with several stages powered by liquid fuel could reach space, thus establishing rocket science. Only in 1926 did the American scientist Robert Goddard [2] design a liquid fuel rocket.

German Contributions

In the mid-1920s, German scientists began to carry out experiments with rockets powered by liquid propellants capable of reaching relatively high altitudes. Wernher von Braun [3] developed such weapons for the Nazis.

The German A-1 missile, launched in 1942, became the first projectile to reach space; in 1943, Germany began production of its successor, the V-2 rocket, with a range of 300 km and a warhead of 1000 kg. The V-2s however were even more deadly to the forced laborers employed in their production at the Mittelbau-Dora concentration camp [4]. The V-2 factory at Mittelbau-Dora produced 4575 missiles between August 1944 and March 1945, the period in which they were launched in batteries (whereas previously they were used for testing and development). It is estimated that of more than 60,000 prisoners employed at the Mittelbau complex, 26,500 died in a 20-month period (estimates of the total number of prisoners vary from 40,000 to 64,000). In particular, 11,000 died when the camps were evacuated by the SS due to the advance of the American army. This evacuation was particularly barbaric: the SS shot the prisoners, crowded them into barns or stables, and burned them alive, leaving them to die if they were too sick to walk or sent them in foot or rail convoys destined for other concentration camps (it was at that period that the Boelcke Kaserne, a barracks in Nordhausen that was later discovered by American troops, became a kind of "dumping ground" for prisoners from various camps who were too sick to be transported. It is little known that the number of

R. G. Russo, *Navigating Health Challenges in Space*,
https://doi.org/10.1007/978-3-032-03620-9_2

people who died from the production of the missiles was greater than the casualties of the missiles themselves. Each operational V-2 from Mittelwerk costed approximately six human lives.

At the end of WWII, US, British, and Soviet soldiers and scientists captured trained personnel from the German missile installation at Peenemünde [5].

The First Space Age (called the space race) is at the same time a chapter of space exploration of the twentieth century and one of the aspects that the Cold War between the United States and the Soviet Union took on, in the period 1957 and 1975.

After World War II, the space race began with the launch of the Soviet satellite Sputnik 1 [6] on October 4, 1957.

NASA was established in the USA (from 10/1/1958: National Aeronautics and Space Administration).

Approximately 4 months after the launch of Sputnik 2, the United States launched its first satellite, Explorer I [7] (January 31, 1958).

Living Creatures in Space

Fruit flies launched by the United States on captured German V-2 rockets in 1946 were the first living things sent into space for scientific study. The dog Laika [8] was the first animal to orbit the earth. In 1960, the Soviet dogs Belka and Strelka orbited the Earth and returned successfully. The American space program sent at least two chimpanzees, and the Soviet Union launched turtles in 1968 to fly around the Moon. Over the past 50 years, American and Soviet scientists have utilized the animal world for testing. Despite losses, these animals have taught the scientists a tremendous amount more than could have been learned without them. Without animal testing in the early days of the human space program, the Soviet and American programs could have suffered great losses of human life. These animals performed a service to their respective countries that no human could or would have performed. They gave their lives and/or their service in the name of technological advancement, paving the way for humanity's many forays into space.

Human Beings

Soviet cosmonaut Yuri Gagarin [9] was the first human to reach space in Earth orbit on the Vostok 1 spacecraft on April 12, 1961.

After Gagarin, the American Alan Shepard entered suborbital space on May 5, 1961, and John Glenn was the first to orbit the Earth, completing three orbits on February 20, 1962.

The Soviet Valentina Vladimirovna Tereškova was the first woman in space on June 16, 1963, followed by Svetlana Evgen'evna Savickaja, who made the first spacewalk, and only in 1983 by the American Sally Ride.

After sending unmanned probes, the lunar landing proceeded. US President John Kennedy and Vice President Johnson proposed the Apollo Program.

President Khrushchev stated in October 1963 that "no flights of cosmonauts to the Moon are being planned." In December 1968, American astronauts James Lovell, Frank Borman, and William Anders [10] orbited the Moon for the first time, as well as celebrated Christmas in space for the first time and returned safely to Earth.

American astronaut Neil Armstrong [10] was the first person to set foot on the Moon on July 20, 1969. Commander of the Apollo 11 mission, Armstrong received support from Command Module Pilot Michael Collins and Lunar Module Pilot Buzz Aldrin in an event followed by more than 500 million people around the world. Armstrong's words were memorable:

> That's one small step for a man, one giant leap for mankind.

Other Successes: Missions to Other Planets

The Soviet Union successfully sent the first probes to Venus and Mars in 1960. The American probe Mariner 10 was the first to reach the planet Mercury in 1974, after having flown by Venus. In 1965, the United States sent the Mariner 4 probe to Mars.

Pioneer 10 flew by Jupiter in 1973 and Pioneer 11 flew by Saturn in 1979, while Voyager 2 was the first probe to fly by Uranus in 1986 and Neptune in 1989.

Conclusion of the Space Race

The Apollo-Soyuz test program of the Apollo and Soyuz probes constitutes the end of the space race.

Development of Space Medicine

In 1981, NASA (National Aeronautics and Space Administration) successfully flew the Space Shuttle for the first time. The Space Shuttle enabled transport of research equipment into space. The Space Shuttle Program enables to take full-fledged equipment for medical experiments into space.

In 1995, Russian cosmonaut Valeri Polyakov, boarding Mir for 438 days, sets the record for the longest stay in space on a single flight. Another Russian cosmonaut, Gennady Padalka, has the record stay in space for 879 days over the course of five flights.

Thus, it was found that humans can live in space for 1 year or more, but the space environment affects individuals differently, and not everyone can safely stay in space for long periods [11]. From the challenges of providing optimal nutrition to managing the risks posed by microgravity, scientists and engineers from NASA work to predict, assess, and solve the problems that humans encounter in space. Through such research, NASA can develop systems that help humans thrive in space. Most astronauts find their freedom from gravity exhilarating, especially as they adapt to their new environment. But weightlessness enormously complicates the business of daily life, from eating to sleeping. And space adaptation involves some very complex changes in the human body, both short-term and long-term.

Space medicine began with the question of whether or not humans can survive in space. Today, the focus is on improving methods of managing astronaut health.

Even the Shortest Trips Have Altered Human Biology

A variety of diverse analyses demonstrate that even a few days in space affects biology in ways that do not appear to be reversible. Not all the changes were reversible after returning to the Earth's surface, even many months after reentry. "We thought they would recover faster than they did."

Genetic changes have been found in how estrogen-related reactions work in space, with greater changes in the B cells of females than males. B cells are related to autoimmune diseases, which tend to affect women more than men.

The conquest of space, or whatever we may wish to call it, is a real change for the way of thinking about our future. When manned space carriers are shot into the universe, man himself can go where up to now only human imagination and its power of abstraction, or human ingenuity and its power of fabrication, could imagine or dream of.

In addition to facilitating our everyday life, space conquest is also essential to the survival of the human species.

Ultimately, space exploration is a necessity with wide-ranging implications. It is not just an endeavor; instead, it involves both intrinsic and extrinsic benefits to humanity.

Bibliography

1. Wikipedia, free encyclopedia, Konstantin Ėduardovič Ciolkovskij.
2. Wikipedia, free encyclopedia, Robert Goddard.
3. Wikipedia, free encyclopedia, Wernher von Braun.
4. Wikipedia, free encyclopedia: Concentration Camp of Mittelbau-Dora.
5. Wikipedia, free encyclopedia, Peenemünde Military Research Center.
6. Wikipedia, free encyclopedia, Sputnik 1.
7. Wikipedia, free encyclopedia, Explorer 1.
8. Wikipedia, free encyclopedia, Laika.
9. Wikipedia, free encyclopedia, Jurij Gagarin.
10. Wikipedia, free encyclopedia, Neil Armstrong.
11. Anderson RW. The cosmic compendium: space medicine. 2015

3 Astronautics (+ Cosmic Energy)

Astronautics is the practice of sending spacecraft beyond Earth's atmosphere into outer space. Spaceflight is one of its main applications, and space science is its overarching field. Astronautics is the theory and practice of human navigation outside the Earth's atmosphere, through the use of both automatic and manned machines. It can also be defined as the science that deals with the technology of space travel.

Psychological and psychiatric problems which occur in space are based on a variety of variables. Astronautics has designed, built, and manufactured best-in-class avionics for over 60 years [1].

What Is the Difference Between Aerospace and Astronautics?

Aerospace engineering is concerned with designing and building aircraft and spacecraft. Aerospace engineering can be divided into two specializations: *aeronautical* engineers, who handle aircraft, and *astronautical* engineers, who design spacecraft, according to the US Bureau of Labor Statistics.

Cosmic Energy Vacuum energy is a quantity of energy present everywhere in space even when devoid of matter, which makes empty space not completely empty. There is abundant energy in space: it is flooded constantly by electromagnetic energy. All stars in the universe produce energy and send it out into space. Therefore, it is an underlying background energy that exists in space throughout the entire universe. Historically, cosmic energy is one of the English translations of the Hindu term *shakti*. It refers to external spiritual energy and can also refer to *prana*, or the life energy that is the source of *kundalini*. Cosmic energy is thought to be a vital source that animates all forms of life and maintains the balance of the entire cosmos.

R. G. Russo, *Navigating Health Challenges in Space*,
https://doi.org/10.1007/978-3-032-03620-9_3

This energy is linked to quantum fluctuations (continuous temporary changes in the energy state of empty space that allow the creation of virtual particle-antiparticle pairs), which determine the continuous fleeting appearance and annihilation of particles and antiparticles [2–4].

The book brings together energy, health, and medicine to try to explain the natural energy that surrounds us in everyday life and how we can maximize its use for our health. Furthermore, it can explain how disease is due to the disagreement between body and mind. The book also explains which allergies must be resolved and not just avoided. Medicines used over the centuries can be accepted and relied on just like all the foods we still consume today. Medicines within the modern allopathic system do not last even a decade; in fact, chemicals, herbs, and minerals are not medicines: they penetrate our body and create a series of side effects and other complications. Only energized medicines (thanks to the use of specific micronutrients such as tonics and energizers) allow us to optimize daily performance and increase our well-being.

Aeronautics and astronautics deal with the design, analysis, and performance of air and space vehicles and a broad spectrum of related engineering science, such as aerodynamics, structural mechanics, automatic controls, flight mechanics, space dynamics, propulsion, plasma dynamics, and related topics.

Space Exploration for the Benefit of Medicine

From the study of prosthetics to telemedicine, there are many innovations developed during NASA space missions that have found application in the medical field.

For instance, what medical advancements came from space?

Here, we look at the top 10 medical breakthroughs made in space, for all mankind:

10. Eye Surgery

Billions of people suffer with some kind of eye disorder, with up to four billion wearing glasses. Many others choose to use contact lenses, while some opt for laser eye surgery to correct their eyesight.

9. Relieving the Symptoms of Menopause

One of the most unpleasant symptoms of menopause is hot flashes, when a woman suddenly feels unbearably and inexplicably hot. Recently, some clothing manufactures designed specific materials for temperature regulation, based on spacesuits.

In the 1980s, NASA worked together with the Triangle Research and Development Corporation to experiment with new materials that would maintain a steady temperature for spacesuits.

8. Improving Balance

Patients suffering from stroke, spinal cord injuries, or motor decline need additional support with their balance—as do astronauts in zero gravity. NASA's preflight and postflight study of eye movement reactions has led to quicker, inexpensive identification and treatment of patients with balance problems. What's more, patients and astronauts can be trained to manage dizziness with a non-pharmacological, computerized method.

7. Shingles

In the USA, over one million people a year are diagnosed with shingles, a condition affecting the nervous system caused by the chickenpox virus.

Scientists from NASA and the University of Colorado teamed up to work together and to find a cure. Together, they detected VZV DNA, which causes chickenpox, in the saliva of astronauts during spaceflight and built a virus detection kit. Now, detection of the virus can be completed in less than one hour and the technology is being used to detect other viruses, too.

6. Amgen

On Earth, bone health can be affected by injury or malnutrition. In space, bone health deteriorates due to loss, which accelerates in the zero-gravity atmosphere with the lack of normal weight-bearing activities human bodies are used to. Astronauts follow a specific nutritional program and follow strict exercises to reduce the impact. Researchers have studied bone loss data carefully, not only to support astronauts in space, but also to aid lives on earth.

Experiments are ongoing to investigate bone loss and muscle atrophy.

5. Ultrasound

Ultrasounds are most commonly used to examine pregnant women and unborn babies, but ultrasound technology is also used to assess various conditions, such as kidney stones, bone fractures, and collapsed lungs.

4. NeuroArm

Canadarm was developed by MDA for the US Space Shuttle Program, which can lift heavy weights and complete complicated maintenance on board the International Space Station. This led to the creation of the neuroArm, a robot that is able to perform surgery within magnetic resonance machines.

The space manufacturing company MDA worked with a team from the University of Calgary to develop the technology to allow surgeons to perform surgeries while a patient was inside an MRI machine, using a robot hand as capable as a human one.

3. Water Purification

Our world is a blue planet from space—but water is one thing space is largely lacking.

Sadly, many humans on earth still lack access to clean water. However, research for ensuring astronauts had a supply of clean water has helped communities on earth, too. The Water Security Corporation teamed up with NASA to use its water filtration and purification system technology, developed for use on space stations, across the world.

2. Vaccines

One thing scientists must look out for prior to spaceflight is the health of the astronauts. If one member of the crew has an illness that spreads to the rest of the passengers during their mission, everyone could be at risk of becoming sick.

Scientists discovered that certain bacteria found in *Salmonella* could become more infectious during spaceflight. *Salmonella* kills over 400 people in the USA each year, as well as causing many hospitalizations. Space scientists researched *Salmonella* and how to combat its deadly bacterial pathogens in confined, low-gravity spaces—which led to a wider understanding of *Salmonella*'s genetic pathway and general microbial vaccine development.

1. Breast Cancer Detection

The second leading cancer-related cause of death in women is breast cancer. Every year, though, treatment for the disease becomes increasingly efficient.

The Image-Guided Autonomous Robot (IGAR) has been refocused inside MRI machines to assist healthcare professionals in the location of breast cancer. IGAR can help increase surgeons' precision during biopsies.

Bibliography

1. Arone A. The burden of space exploration on the mental health of astronauts: a narrative review. Clin Neuropsychiatry. 2021;8(5):237.
2. Carroll RW. Fluctuations, information, gravity and the quantum potential. Amsterdam: Springer; 2006.
3. Battersby S. It's confirmed: matter is merely vacuum fluctuations. New Sci. Retrieved-06-18, 2020.
4. Ko Paandu, Cosmic energy and the nature's way in health and medicine, Paperback – September 29, 2015.

Part II

Medicine in Space

4 Physiology and Microgravity

Space physiology is concerned with the structure and functioning of the body under conditions encountered by space travelers; these conditions have mostly been confined to the environment of the spacecraft that houses the astronauts. In the future, however, space physiology will include the atmospheric and gravitational conditions found on planets, moons, or other stellar bodies. But even orbiting around Earth poses difficulties for the astronauts. The reduced gravity of a spacecraft makes it difficult for the body to distinguish "up" from "down." On Earth, such distinction by the vestibular organ of the inner ear is easy, because of the orienting power of gravity.

Previous studies have demonstrated that long-term exposure to microgravity during spaceflight can cause various changes in the human body, such as muscle atrophy, changes in muscle fiber composition, changes in cardiovascular function, decreased bone density, and even disruptions in immune system functions.

The main factors that can cause biological changes and serve as study targets on the International Space Station (ISS) include radiation, confinement and isolation, extreme temperatures, circadian disruptions, and microgravity. Astronauts experience apparent weightlessness in space due to microgravity, a condition where inertial forces balance the force of gravity. This sickness is usually transient, and astronauts acclimatize soon after going into orbit.

The longer missions of the future will need to incorporate more Earth-like gravitation conditions or an exercise regimen, or both.

The International Space Station (ISS), orbiting Earth at an altitude of 330–410 km, is within a gravitational field that exerts 88.9% of the gravity felt on Earth's surface. However, its constant velocity of 27,600 km/h results in free fall, leading to the sensation of weightlessness for astronauts. The physical and biological changes observed in space highlight the significant influence of gravity on human evolution on Earth. The impacts of microgravity on the human body can be divided into two primary categories: effects that occur after short-term exposure, primarily

R. G. Russo, *Navigating Health Challenges in Space*,
https://doi.org/10.1007/978-3-032-03620-9_4

consisting of space adaptation syndrome (SAS) and temporary nausea caused by an imbalance of the vestibular system, and effects that occur after long-term exposure [1].

Astrophysics

Astrophysics is a blend of physics and astronomy, aiming to understand the environment surrounding our planet. This includes diverse phenomena from high-energy charged particles in the plasma around the Earth to the explosive release of magnetic energy from the Sun. Without the understanding of the universe that physics gives us, space science would not be possible at all—no physics, no space science.

Space studies in humans and animals have provided only a snapshot into understanding the role of gravity on physiological responses. Fully understanding this relationship, including adaptive mechanisms, will provide the information required to ensure normal physiological function in crew for long-duration space missions.

The US has mainly concentrated on relatively short flights. The Russian studies provided information about long-term exposure to microgravity. The Soviet/Russian program concentrated on first the Salyut and then the Mir space stations. The most important physiological changes caused by microgravity include bone demineralization, skeletal muscle atrophy, vestibular problems causing space motion sickness, cardiovascular problems resulting in postflight orthostatic intolerance, and reductions in plasma volume and red cell mass.

Microgravity is the condition in which people or objects appear to be weightless. The effects of microgravity can be seen when astronauts and objects float in space. Microgravity can be experienced in other ways, as well. "Micro-" means "very small," so microgravity refers to the condition where gravity seems to be very small. In microgravity, astronauts can float in their spacecraft—or outside, on a spacewalk. Heavy objects move around easily. For example, astronauts can move equipment weighing hundreds of pounds with their fingertips. Microgravity is sometimes called "zero gravity," but this is misleading.

The first part of this book [2] summarizes the fundaments of transport theory relevant to the atmosphere, ionosphere, and magnetosphere. This theory is then applied to physical phenomena in space. Under the heading "astrophysics," we identify a huge field of research that has taken on its own vast, more precise physiognomy, due to the possibility of direct exploration of space around Earth and, more generally, of interplanetary space. In this context, physics has a distinctly interdisciplinary character covering and combining—among others—plasma physics, solar physics, planetary physics, and geophysics.

The proprioceptive system—the system of nerves in the joints and muscles that tell us where the arms and legs are without any visual inspection—can also be affected. Low gravity reduces or eliminates the tensions impinging on the joints and muscles, which can make the appendages appear invisible to the brain. Such confusion between what the eye sees and the brain perceives can result in what has been termed space sickness. Space sickness is usually transient, and astronauts acclimate soon after going into orbit. Microgravity also affects the skeletal structure of astronauts. The absence of stress-bearing activity and the loss of components of the bones, particularly calcium, have produced shortening and weakening of bones and the atrophy of muscles in astronauts who have been orbiting the Earth for just several months. Even the heart becomes smaller. The extended missions of the future will need to incorporate more Earth-like gravitation conditions or an exercise regimen, or both.

Why Does NASA Study Microgravity?

NASA studies microgravity to learn what happens to people and equipment in space. Microgravity affects the human body in several ways. For example, muscles and bones can become weaker without gravity, making them work as hard. Astronauts who live on the space station spend months in microgravity. Astronauts who travel to Mars also would spend months in microgravity, traveling to and from the Red Planet. NASA must learn about the effects of microgravity to keep astronauts safe and healthy. In addition, many things seem to act differently in microgravity. Fire burns differently. Without the pull of gravity, flames are more round. Crystals grow better. Without gravity, their shapes are more perfect. NASA performs science experiments in microgravity. These experiments help NASA learn things that would be hard or perhaps impossible to learn on Earth.

A microgravity environment leads to changes in fluid distribution, muscle loading , and altered signaling. Some basic changes include alterations in blood pressure and the quantity of blood that is pumped by the heart with each beat.

Space physiology tries to characterize body responses to space, especially weightlessness, reduced activity, and stress. It provides the necessary knowledge required for an efficient space medicine. Cardiac output, stroke volume, pulse pressure, and heart rate significantly increased with the gravity level [3]. These results confirm that a combination of exercise and artificial gravity provides greater cardiovascular benefit than exercise alone.

Bibliography

1. Manna OM, Stefano Burgio S, Picone D, et al. Microgravity and human body: unraveling the potential role of heat-shock proteins in spaceflight and future space missions. Biology. 2024;13(11):921.
2. Gombosi TI. Physics of the space environment, Edit. Cambridge University Press, 2004.
3. Gilles C, Boyle RD, Gunga H-C. Editorial: The effects of altered gravity on physiology. Front Physiol. 2019;10:1447.

5 Astrophysics and Astrochemistry

Space physics is a blend of physics and astronomy that aims to understand the environment surrounding our planet. This includes diverse phenomena from high-energy charged particles in the plasma surrounding the Earth to the explosive release of magnetic energy from the Sun. Without the understanding of the universe physics gives us, space science would not be possible at all. No physics, no space science.

Under the heading "astrophysics," a huge field of research is identified which has taken on its own vast and more precise physiognomy, due to the possibility of direct exploration of the space surrounding the Earth and more generally of interplanetary space. Since the functions of the human body are not normal in space, it is improper to talk about physiology: in such cases, it is more correct to talk about adaptation.

As such, physics has a distinctly interdisciplinary character, as, among others, plasma physics, solar physics, planetary physics, and geophysics coexist within it.

The space surrounding the Earth and the planets and, in general, the interplanetary space constitute an immense laboratory in which electrically charged particles, magnetic and electric fields, electromagnetic radiation, and collective thermodynamic and electromagnetic phenomena manifest themselves and influence each other.

Astrophysics must be interpreted in the sense of physics ("in" space), that is, the study and interpretation of phenomena that occur in space. Some basic topics are included, namely the solar wind and its interaction with the external envelope of the planets, up to the limits of interplanetary space, as well as the formation of the magnetosphere and radiation belts [1].

The Center for Astrophysics | Harvard and Smithsonian is involved in researching medical applications. Indeed, some of the same challenges arise in astronomy and medicine, despite the very different focus of each field. Both disciplines need to produce high-quality images of systems that are often difficult to study, albeit for very different reasons. Both require the use of types of radiation that are often difficult to focus and harmful to equipment. For this reason, detector technology and

R. G. Russo, *Navigating Health Challenges in Space*,
https://doi.org/10.1007/978-3-032-03620-9_5

data techniques developed for astronomy have found fruitful application in medical research.

The Solar Wind [2] The solar origin of the electromagnetic radiation that reaches Earth is well known. About half of the energy conveyed by this radiation is found in the visible region of the spectrum. This, however, extends from the lowest frequencies, lower than those of radio waves, to the much higher ones (e.g., X and γ rays). That the Sun was also a source of corpuscular radiation is a fact that has long been postulated on the basis of indirect arguments, but it has only recently been verified experimentally using instruments carried by spacecraft in interplanetary space.

The idea that the zodiacal light, a weak luminosity observable near the equatorial plane, was light diffused by corpuscles dates back to G. D. Cassini (1672) [3]. The theory of S. Chapman and V. C. A. Ferraro [4] finally took shape, attributing magnetic storms and polar auroras to the interaction of solar particle beams with the Earth's magnetic field.

The initial theory was that charged particles originated from the Sun sporadically; the idea subsequently suggested was that the Sun emits particles continuously, that is, it is a permanent source of corpuscular radiation. From a theoretical point of view, the general properties of this radiation were predicted by E. N. Parker in 1957 who then coined the term "solar wind."

The first direct observations of the solar wind were made by K. Gringauz and collaborators in 1959 and subsequently by the American satellite Explorer. The fact that the particles are electrically charged, mainly protons and electrons, together with the observation that the Sun has intense magnetic fields, implies that the solar wind is also permeated by a magnetic field, i.e., a magnetized plasma.

"Contact" with the Sun can be good for the skin. The Sun's rays, amplifying the effects of trace elements found in seawater, stimulate cell turnover, almost producing a peeling effect and promoting wound healing. Sunrays are useful for psoriasis, as they reduce the thickness and redness of the typical patches. Even dermatitis, both atopic (red patches, itching, blisters, dry skin) and seborrheic (red patches, greasy scales), improves thanks to the antiseptic action of the Sun.

Astrochemistry

Astrochemistry, a branch of chemistry, is the interdisciplinary scientific study between chemistry, astronomy, and physics; it studies the chemical elements present in space, typically in clouds of molecular gas, their formation, growth, interaction, and destruction. In practice, it is a fusion between astronomy and traditional chemistry. Astrochemistry often requires the use of telescopes to make measurements on various aspects of celestial bodies, such as their temperature or chemical composition.

Astrochemistry partially overlaps astrophysics in the study of nuclear reactions in the stars, the consequences of stellar evolution, and the formation of elements.

Radio astronomy was already developed in 1930, but it was only in 1937 that an interstellar molecule was first identified. Until then, the only known species in interstellar space were atomic. The discovery was confirmed in 1940, when the molecules of CH and CN were assigned the spectroscopic lines still unidentified at that time, but detected in space.

In the next 30 years, additional molecules were found including OH, discovered in 1963 and considered an important indicator of oxygen in interstellar space, and H_2CO (formaldehyde), discovered in 1969—the first polyatomic molecule found in space.

The astrochemistry of the vast molecular clouds opened the way to astrobiology, and the search for life in the universe.

There is a scientific study describing how galaxies of stars depend on the evolution of chemistry [5]. These discoveries aim to inform the chemical sciences and engineering communities.

Throughout much of human history, space was thought to be a void in which only ions or radicals existed. It was only in the last half of the twentieth century that scientists began to discover the existence of molecules, such as ammonia, in space. These exciting discoveries in astrochemistry, a multidisciplinary field that focuses on the chemical composition of and processes in astrophysical and planetary environments, have potential applications to the general chemistry and chemical engineering communities. Accordingly, the National Academies of Sciences, Engineering, and Medicine held a workshop on November 8–9, 2018, to (1) explore the chemistry of space, its novel chemicals and reaction mechanisms; (2) discuss information from remote sensing through spectroscopy; and (3) consider discoveries from spacecraft missions in the solar system and laboratory studies of extraterrestrial samples. The discipline is an overlap of astronomy and chemistry.

Bibliography

1. Gombosi TI. Physics of the space environment, Edit.: Cambridge University Press, 2004.
2. Treccani Italian encyclopedia, Physics, 1981.
3. The solar wind is a flow of charged particles emitted from the Sun's upper atmosphere: it is generated by the continuous expansion of the solar corona in interplanetary space. This flow is mainly composed of electrons and protons with energies normally between 1.5 and 10 keV.
4. Giovanni Domenico Cassini (Perinaldo, 8 June 1625 – Paris, 14 September 1712) was a mathematician and astronomer. Italian naturalized French: engineer, doctor, and biologist. He was a professor of astronomy at the University of Bologna, and in 1671, he became the director of the Paris Observatory. He was the father of Jacques Cassini, also an astronomer and grandfather of César-François Cassini, astronomer and geodesist. He died in Paris on 14 September 1712 at the age of 87.
5. Zahid J H, Geller MJ, et al. Astrophys J Lett. 2013;771(2):article id. L19, 6 pp.

Biochemistry and Pharmacology

6

This book introduces a paradigm shift in biochemistry [2], yet a critical oversight persists in neglecting the higher dimensionality of molecules and its profound implications for living organisms.

Eliminating gravity's influence alters biological systems, enabling experiments that cannot be done on the ground. Researchers are sending materials into space to study treatments for cancer, heart diseases, neurological disorders, blindness, and other conditions. Such investigations extend beyond civilian medicine.

Scientists from around the world are studying the effects of microgravity and cosmic radiation via the "off-Earth" International Space Station (ISS) laboratory platform. The ISS has helped scientists make discoveries that go beyond the basic understanding of Earth. Over 300 medical experiments have been performed to date, with the goal of extending the knowledge gained for the benefit of humanity. The medical contents are trifold structured, starting with the well-being of space travelers (astronaut health studies), followed by medical formulation research under space conditions, and then concluding with a blueprint for space pharmaceutical manufacturing. The review covers essential elements of our Earth-based pharmaceutical research such as drug discovery, drug and formulation stability, drug–organ interaction, drug disintegration/bioavailability/pharmacokinetics, pathogen virulence, genome mutation, and the body's resistance. The information compiles clinical, medicinal, biological, and chemical research as well as fundamentals and practical applications.

NASA's innovations have led to a variety of medical devices and procedures, including ear thermometers, insulin pumps, artificial heart defibrillators, and improvements in mammography technology, demonstrating the broad impact of space technology on healthcare (Mar. 7, 2024).

Space biomedicine has provided significant technological breakthroughs by developing new medical devices and diagnostic tools [3].

R. G. Russo, *Navigating Health Challenges in Space*,
https://doi.org/10.1007/978-3-032-03620-9_6

Microorganisms

Wherever there are humans, there are also bacteria. This is perhaps how the microorganisms that evolved with us on Earth have also "conquered" space and are adapting. This is confirmed by the research of scientists at NASA's Jet Propulsion Laboratory, based at the California Institute of Technology, who have been collecting and analyzing samples taken from the ISS for about a decade to monitor changes and ensure that there are no threats to astronauts in orbit.

The researchers isolated many bacterial species from samples taken in different environments of the ISS and compared them with species that also exist on Earth. It was thus noted that microbes that lived in low Earth orbit are differentiating, with adaptations that help them survive in extreme conditions. In fact, it turns out that over 250 different species of bacteria and fungi can survive in outer space. Even more shocking, they actually thrive there.

Bacteria on the International Space Station have transformed into something never seen before; an international research team has observed that bacteria isolated on the ISS have developed unique genetic, metabolic, and functional characteristics, never seen before on Earth. They have adapted to life in space by modifying some proteins to make them more functional in microgravity conditions, and their DNA repair system is more active to counteract the effects of increased exposure to radiation. Furthermore, scientists observed the presence of mobile genetic elements that improved their metabolism.

They are a risk because they are super resistant to drugs.

Microorganisms have, in fact, developed unique genetic, metabolic, and functional characteristics, never seen before. According to experts, this radical divergence is linked to the extreme environment of the orbiting laboratory, where bacteria must face multiple challenges—among them microgravity, high levels of radiation, and CO_2 (carbon dioxide).

In the latest study, Kasthuri Venkateswaran and colleagues studied newly discovered bacterial species found in ISS samples: *Microbacterium mcarthurae*, *Microbacterium meiriae*, *Paenibacillus vandeheii*, *Arthtobacter burdickii*, and *Leifsonia williamsii* (M.Maestroni, Sept. 20, 2024). Researchers also found land and marine bacteria in cosmic dust samples collected during a sidewalk.

Moreover, NASA discovered 13 strains of a superbug, a multidrug-resistant bacterium. Its name? *Enterobacter bugandensis* (Jul. 21, 2024). The discovery is significant because that bacteria, notoriously resistant to antibiotics, is a superbug classified as an opportunistic pathogen, which is capable of triggering infections in "favorable" conditions, for example, in sick, immunosuppressed, and immunocompromised people.

In response to the question of whether astronauts left bacteria on the Moon, it is possible to answer that as part of the Apollo 12 mission in November 1969, the camera from the Surveyor 3 probe was brought back from the Moon to Earth. Upon analyzing the camera, it was found that the common bacterium *Streptococcus mitis* was alive on the camera.

It is no wonder that bacteria brought by humans into space can evolve in a totally different way than what is seen on Earth.

Space travel is on the rise, with longer and longer durations. Interest in astropharmacy (pharmacology in space), which studies the effects of space travel on the effects of drugs, is growing.

Are mutant space bacteria a threat?

What particularly arouses the attention of experts, however, is the fact that genetic traits linked to pathogenic potential have emerged in the bacterial species living on the ISS: some genes associated with virulence (such as those that help evade or damage the immune system) are more active. Furthermore, it seems that space microbes are able to form biofilms on the surfaces of the ISS—an ability that makes them more resistant to disinfectants and antibiotics.

Whether this represents a concrete risk for the health of astronauts to date is not yet clear (there is, however, no alarm), but it is confirmed that studying the adaptations of microbes to space is of extreme importance in anticipation of long-term space travel. First of all, we are now aware that it is necessary to implement more incisive actions to prevent the proliferation of bacteria on surfaces (for example, better control of the humidity of the environments on the ISS); then, those same mutations could become targets for new drugs, if the modified bacteria prove to be a health risk.

But—the scientists conclude—we could also discover that these superbugs are a resource.

A different characteristic concerns the bacterium *Deinococcus radiodurans* [2]. This extraordinarily resistant bacterium withstands harsh environmental conditions present in outer space. In fact, this bacterium is an extremophile and one of the most radiation-resistant organisms known. It can survive cold, dehydration, vacuum, and acid and therefore is known as a polyextremophile. When exposed for 1 year outside the ISS within the Tanpopo orbital mission to investigate microbial survival, it was the first unknown organism to be sequenced and identified entirely in space (Jan. 2, 2018) [4].

Researchers also found land and marine bacteria in cosmic dust samples collected during a spacewalk (Jan. 19, 2024).

Bibliography

1. AksoyalpZ Ş, Temel A, Karpuz M. Pharmacological innovations in space: challenges and future perspectives. Pharm Res. 2024.
2. Zhizhin GV. Biochemistry in the space of the highest dimension. IGI Global; 2024.
3. Zahid JH, Geller MJ, et al. The chemical evolution of star-forming galaxies over the last 11 billion years. Astrophys J Lett. 2013;771(2) article id. L19, 6 pp.
4. Bizzarri M, Fedeli V, et al. Space biomedicine: a unique opportunity to rethink the relationships between physics and biology. Biomedicines. 2022;10(10):2633. Pub. Online.

Drugs and Astropharmacy 7

Astropharmacy research helps make human space exploration safe and accessible. Astropharmacy is concerned with the development of drugs and medical treatments that can be used in microgravity conditions and evaluates their safety in space.

In space, drugs are used to treat both common symptoms and emergency situations such as trauma and bleeding. In particular, medical kits provided to astronauts include drugs to treat sleep disorders, allergies, space motion sickness, nausea, pain, and sinus congestion.

The most commonly used drugs in approximately 20 space flights lasting more than 30 days were zolpidem and zaleplon for the treatment of insomnia and ibuprofen, paracetamol, and aspirin for the treatment of joint and bone-muscle pain and headaches. In addition, topical preparations based on corticosteroids, antihistamines, and anti-infectives are used to prevent or treat conditions such as skin rashes. Finally, the intake of vitamin D, calcium, and bisphosphonates has proven useful.

Space flight involves alterations in human physiology resulting in changes in pharmacokinetics, such as drug absorption, distribution, metabolism, and excretion, and pharmacodynamics. All of these can inevitably impact the efficacy and safety of drugs.

The efficacy and safety of drugs in space also depend on their stability and preservation. Space missions present unique environmental and physical factors, including microgravity, intense vibration and radiation, and variable humidity and temperature, which can significantly contribute to the instability of pharmaceutical products.

Space pharmacology is the application of biomedical engineering that studies the use and dynamics of drugs or pharmaceuticals in space environments.

The effect of outer space on the subject has been a recent concern on drug delivery and mechanism. There is a need to modify drug doses and release profiles to achieve maximum drug efficacy in outer space due to environmental implications on the human body.

R. G. Russo, *Navigating Health Challenges in Space*,
https://doi.org/10.1007/978-3-032-03620-9_7

Pharmacological studies can determine the effects of chemical agents upon subcellular, systemic, physiological, or behavioral processes; focus on the treatment and prevention of diseases; or deal with the potential hazards of pesticides and herbicides [1].

What Is the Importance of Astropharmacy?

Research in astropharmacy is helping make human exploration in space safe and accessible. Astropharmacy deals with the development of drugs and medical treatments that can be used in microgravity conditions and evaluates their safety in space.

What Are the Most Important and Most Widely Used Categories of Drugs During a Manned Space Mission?

In space, drugs are used both for the treatment of symptoms that may occur commonly and for any emergency situations such as trauma and bleeding. In particular, medical kits provided to astronauts include medications to treat sleep disorders, allergies, space motion sickness, nausea, pain, and sinus congestion.

Are There Any Examples of the Use of These Drugs?

The drugs most used by 24 US crew members on 20 spaceflights lasting more than 30 days were zolpidem and zaleplon for the treatment of insomnia and ibuprofen, acetaminophen, and aspirin for the treatment of joint and musculoskeletal pain and headaches. In addition, astronauts use topical corticosteroids, antihistamines, and anti-infectives to prevent or treat conditions such as rashes. Finally, for astronauts who spend long periods in space, with loss of up to 10% of bone mineral density, it is useful to take vitamin D, calcium, and bisphosphonates.

The pharmacokinetics and pharmacodynamics of drugs change in space, as do absorption, distribution, metabolism, and excretion. All of these can inevitably have an impact on the efficacy and safety of medicines.

Space pharmacology is the application of biomedical engineering to study the use and dynamics of drugs or pharmaceuticals in space. The effects of outer space on an astronaut have aroused recent interest in drug delivery and mechanisms.

The effects of outer space on the human subject have recently aroused concern about drug delivery and mechanisms [2].

Pharmacological studies can determine the effects of chemical agents on subcellular, systemic, physiological, or behavioral processes; focus on the prevention and treatment of diseases; or deal with the potential hazards of pesticides and herbicides [3].

For orally administered drugs, absorption may be modified by delayed gastric emptying and slowed intestinal transit caused by microgravity conditions, spatial motion sickness, and interaction with drugs used to manage nausea, resulting in decreased plasma levels of the drug.

On the other hand, the movement of body fluids towards the head and the decrease in plasma volume can affect the volume of distribution of drugs taken, potentially modifying their clinical response.

The same reasons apply to metabolism. Changes in blood flow can alter hepatic metabolism and liver biotransformation enzymatic activity, with the risk of significantly raising or lowering blood concentrations of the drug. These changes can result in increased toxicity or reduced efficacy of the drug, respectively.

How is drug elimination affected? This is where microgravity, nutritional changes, and dehydration come in, which can affect urinary drug excretion. Microgravity, nutritional changes, and dehydration can cause changes in urinary drug excretion.

Having the right medicines onboard is crucial for long stints in space, which can last months at a time while complex tests and experiments are carried out in the microgravity environment. Therefore, we can expect that the efficacy and safety of drugs in space also depend on their stability and preservation. Space missions present unique environmental and physical factors, including microgravity, vibration and intense radiation, and varying humidity and temperature, which can significantly contribute to the instability of pharmaceuticals.

The potential risks of drug instability in space include the following: (a) chemical instability, resulting from undesirable chemical reactions, which may cause the active ingredients to lose efficacy, may alter excipients or cause the formation of toxic degradation products; (b) physical instability manifested by changes in the appearance, viscosity, shape, organoleptic properties, and pH of drugs, which could compromise their safety.

What Drugs Are Used in Space? [4]

Benzodiazepines are frequently used medications in space, though less often than non-benzodiazepine "z-drugs." The long acting nature of benzodiazepines used by astronauts, such as temazepam, have been labeled "non-ideal" for spaceflight use due to a high tendency of causing morning impairments.

Non-benzodiazepines sedative-hypnotics (also known as "z drugs") such as zolpidem, zopiclone, and zaleplon are the most frequently used medications on shuttle missions. However, despite their widespread use among astronauts, there has been relatively little research on non-benzodiazepines in spaceflight [5].

Space Medicine is a subspecialty of emergency medicine which evolved from the aerospace medicine specialty. It is dedicated to the prevention and treatment of medical conditions that would limit success in space operations. Moreover, it focuses specifically on prevention, acute care, emergency medicine, wilderness medicine, and hyper/hypobaric medicine in order to provide medical care to astronauts and spaceflight participants. Astronautical hygiene is the application of science and technology to the prevention or control of exposure to the hazards that may cause illness in astronauts [6].

Medical Contingency on ISS

Quite a few unexpected medical conditions are possible on ISS, due to its prolonged presence in space. NASA provided resources to handle in-flight medical contingencies. Medical emergencies during space flight might include arrhythmias, heart attacks, strokes, embolisms, hemorrhages, renal stone formations, infections, or even thrombotic complications. As a matter of fact, a few of these emergencies have been documented during space flights.

There may be the eventuality that these cases must be treated on orbit, with return to Earth delayed or avoided. Of course, should this happen, numerous changes must be completed including modifications to training practices, treatment guidelines, diagnostic and therapeutic resources, and also informatics.

Moreover, a defibrillator, a portable ultrasound, a device that can look deep into the eye, and two liters of saline are present on ISS.

In addition, NASA also maintains a robust pharmacy and a suite of medical equipment onboard the space station to treat various conditions and injuries. Of course if medical emergency requires it, a return to Earth might be considered.

Bibliography

1. Tran QD, Tran V, et al. Space medicines for space health. ACS Med Chem Lett. 2022;13(8):1231–47.
2. Krisko A, Radman M. Biology of extreme radiation resistance: the way of Deinococcus radiodurans. Cold Spring Harb Perspect Biol. 2013;5(7):a012765.
3. Wotring VE. Space pharmacology. Springer International Publishing AG; 2023.
4. Aziz S, Raza MA, Noreen M, Iqbal MZ, Raza SM. Astropharmacy: roles of pharmacist in space. Innov Pharma. 2022;13(3)
5. Hodkinson PD, Anderton RA, Posselt BN, Fong KJ. An overview of space medicine. Br J Anaesth. 2017;119(suppl_1):i143–53.
6. Garrett-Bakelman FE, Darshi M, et al. The NASA Twins Study: a multidimensional analysis of a year-long human spaceflight. Science. 2019;364(6436)
7. Khedekar D. Advancing clinical pharmacology through microgravity: insights and framework from space-based research programs. J Pharma Exp Ther. 2024;389(S3):061.

The Human Anatomy in Space: Effects on Health

8

How the Human Body Changes in Space

The changes that the space environment causes on the human body have been studied and documented for decades [1]. The best known are bone decalcification and fragility, muscle loss, reduction (temporary or prolonged) of sensory and motor capabilities, and cardiovascular and ocular disorders.

The main question is, "How much does the biology of astronauts change after carrying out a space mission?" A plausible answer is that "Humanity is far from an interplanetary species, but this is the first of many steps in preparing for the Moon and Mars," says Christopher Mason of Weill Cornell Medicine, a biophysicist and physician.

The changes that the space environment causes on the human body have been widely studied [2]. Hence, to get an overview of all the changes an astronaut's body undergoes in space, NASA sent Scott into space for 340 days, while his twin brother Mark remained on Earth. The first important evidence is bone decalcification and fragility, muscle loss, reduction (temporary or prolonged) of sensory and motor skills, and cardiovascular and ocular disorders. In the majority of cases, these disorders can be reduced with adequate physical activity during the stay in space and gradually disappear when the astronaut returns to Earth. The first relevant evidences are bone decalcification and fragility, muscle loss, reduction (temporary or prolonged) of sensory and motor skills, and cardiovascular and ocular disorders. In the vast majority of cases, these disorders can be reduced with adequate physical activity during the stay in space and gradually disappear when the astronaut returns to Earth.

Because they have the same genetic makeup, twins are ideal for analyzing how the environment influences human health, regardless of changes related to genetic predisposition.

Scott Kelly boarded the International Space Station in March 2015 and stayed there for 340 days. During this time, his brother Mark, who remained on Earth, was

R. G. Russo, *Navigating Health Challenges in Space*,
https://doi.org/10.1007/978-3-032-03620-9_8

subjected to the same battery of tests as his twin. Physiological, telomeric, transcriptomic, epigenetic, proteomic, metabolic, immune, microbiomic, cardiovascular, ocular and cognitive data were collected for 25 months (before, during, and after Scott's flight).

This periodic check was published in April 2019. According to Susan Bailey, co-author of the article, it is the "most comprehensive assessment we have ever had of the human body's response to space flight."

Some functions remained similar in the two brothers; others differed significantly but returned identical after the flight. Scott's telomere length increased significantly during his time in space, returning to "normal" a few days after his return to Earth. There were some changes in gene expression. Scott's gut microbiome clearly changed during the mission, returning to normal at the end of the study.

Temporary Effects

Researchers believe that some of the DNA damage is attributable to the body's exposure to radiation. Cognitive abilities remained unchanged (e.g., mental alertness and spatial orientation). The functions of the human body in an alien environment such as space significantly change, from the organ level to DNA. Almost all these changes disappear upon returning to Earth.

Below is a concise summary of the main changes of the body in space:

Bones

Without the need to carry the load of Earth's gravity, the bones in the lower part of your body (legs, hips, and spine) are doing less work. The decrease in load bearing results in your bones losing their density. More problems appear like increased urinary calcium and renal stones.

Muscles

Astronauts experience decreased muscle mass, strength, and endurance because moving around requires reduced work from the legs and back. The muscles can begin to weaken or atrophy.

Neurological

Your brain has a complex, integrated set of neural circuits that allow us to maintain balance, stabilize vision, and orient your location and direction. Lacking gravity, the data inputs from muscles and internal organs are different from what is expected. Early in space missions, astronauts can experience disorientation, space motion sickness, and a loss of sense of direction. On returning to Earth, the astronauts are immediately forced to readjust again, back to Earth's gravity.

Eyes

The eyes also experience alterations when in space, especially during long-duration missions. Changes include increased globe flattening, optic disc edema, and vision changes. One of the most recognized issues for the eyes in space is space-associated neuro-ocular syndrome (SANS), which causes swelling in the back of the eyes.

Cardiovascular

In space, astronauts may face decreases in blood volume, orthostatic tolerance, and aerobic capacity while also experiencing increased arrhythmias. Over time, this could lead to deconditioning and a decrease in the size of the heart. On return to Earth and gravity, a smaller and weaker heart could overextend itself. Finally, there is also a concern that space radiation may affect endothelial cells in the lining of blood vessels, which might initiate or accelerate coronary heart disease.

Performance

Factors like being confined in a small space with other people, high work demands, team and crew issues, the loss of the 24-hour day/light cycle leading to fatigue, and stress can all lead to decreased cognitive function. These could pile up to jeopardize a mission. Scientists are studying to protect astronaut bodies and minds to preserve their performance in space.

Gastrointestinal System

The gastrointestinal system feels the effects of a loss of gravity. Without gravity to help move food through your GI tract, the intestinal system can decrease motility.

Environmental Effects

Noise exposure during launch, mid-flight, and landing can cause hearing loss. Not standing on your feet causes them to lose callouses and become soft from disuse and less sensitive to pressure variations. Astronauts on the ISS, have developed callouses on top of their feet from using footholds on the station. Space explorers may also experience skin irritation from microbial growth combined with a lowered immune system.

Space Radiation

While on Earth, the atmosphere protects humans from harmful space radiation. Beyond low Earth orbit, space radiation can pose a significant risk to space explorers. They can experience radiation sickness, central nervous system effects, degenerative diseases, and most notably, an increased lifetime risk of cancer. On Earth, we are exposed to a relatively minimal amount of radiation on any given day—but in space, astronauts are exposed to about 100 times that. That could be anywhere from 150 to 6000 chest X-rays.

Because they share the same genetic makeup, twin studies allow us to analyze how human health is influenced by the environment, regardless of physical changes related to genetic predisposition.

Scott boarded the International Space Station in March 2015 and stayed there for 340 days. During this time, his brother Mark, who remained on Earth, was subjected to the same battery of tests as his twin. Physiological, telomeric, transcriptomic, epigenetic, proteomic, metabolic, immune, microbiomic, cardiovascular, ocular and cognitive data were collected for 25 months (before, during, and after Scott Kelly's flight).

This periodic check was published in April 2019. According to Susan Bailey, co-author of the article, it is the "most comprehensive assessment we have ever had of the human body's response to spaceflight."

Some functions remained similar in the two brothers; others differed significantly, but returned identical after the flight. There were some changes in gene expression..

Bibliography

1. Formanek AR III. Anesthesiology and critical care medicine in space: a clear a4d present need. ASA Monitor. 2018;82:52–3.
2. Jain V. Microgravity, international journal of gynecology & obstetrics, and developmental cell. Noa Leach; 2023.

9 Anesthesiology and Gynecology in Space

Anesthesiology in Space

Astronauts are subjected to copious radiation, neurological feedback loops are altered, and fluid shifts become readily apparent. Removal of a downward venous gradient over time is now thought to lead to intracranial hypertension with consequences resembling those of uncontrolled pseudotumor cerebri for long-duration space missions. Breathing circuits in the spacecraft must be managed to allow for adequate fresh gas flow and CO_2 removal while still maintaining allowable engineering parameters. As experts in hemodynamics, breathing mechanics, and gas exchange, anesthesiologists are uniquely qualified to understand and manage the complex changes occurring to the human body during spaceflight. These and new challenges will only become more apparent as human space exploration continues to expand and private companies begin offering commercial spaceflight.

Earth's First Space Gynecologist Explains How Humans Will Reproduce in the Future

Future human races may live extraterrestrial lives on a variety of newly discovered planets. But how will we reproduce and continue survival once we are up there? What happens to our reproductive systems? Are babies in space even possible? To find out, we spoke to space gynecologist, Dr. Varsha Jain from the University of Edinburgh, about women in space and reproductive health.

R. G. Russo, *Navigating Health Challenges in Space*,
https://doi.org/10.1007/978-3-032-03620-9_9

What Does It Actually Mean to Be a Space Gynecologist?

It's a title that was given to me a few years ago, and I quite like it! I think it stems from the fact that I wanted to take an academic approach to women's health related to astronaut health: I started doing that about 10 years ago. The reason is that no one knows what would happen to women's periods when they go into space. There was a big concern that, instead of coming out of the body, the blood would flow inside the body: a process called retrograde menstruation. There was also a big conversation around how many tampons astronauts would need.

What Does Happen to Our Reproductive Systems in Space?

It's important to say that astronauts are in space to do their job—meaning they are not research participants. We've got to accept that when we receive information and data from astronauts they're volunteering that information.

It's also a very small pool for us researchers to work with: in all, about 650 people in US and European data have been into space. Only 10% or 11% of those are female—so it's not characteristic of the Earth's population.

That said, what we do know is that our reproductive health is governed by our hormones, and the hormone cycles don't appear to change in space. That means menstruation if there's not a pregnancy—and, theoretically, if there was a conception, potentially a normal pregnancy.

Is It Possible to Conceive, Get Pregnant, and Give Birth to Babies in Space?

Because we know women can have their periods in space, it's definitely possible for ovulation to occur. This means that the lining of the womb has grown, so it's preparing for pregnancy. Therefore, technically fertilization might occur as well.

Whether that's actually possible is a real unknown. I'm not privy to that information, if that sort of research is even being done.

Right now, there's a focus on how we can keep astronauts healthy. I think that needs to be a priority rather than finding out if we can have babies in space.

The effects of space on pregnancy have been studied in animals, but not yet on human models. In animals, we know that the pregnancies can continue and deliver because rodent pups have been delivered in space.

What Do We Know About Risks to Women's Bodies and Pregnancies in Space?

A woman on Earth would be advised not to get an X-ray because of the potential impact on the baby. On a mission to space, however, we don't know how the background radiation will impact things—and we don't want to cause a negative impact to an unborn child.

Radiation impacts our genetic makeup—the code that determines who we are. When you have excessive amounts of radiation, it can change that code and can cause mutations. A fetus's genetic code is very susceptible to radiation and very liable to damage. This could lead to differences in how the child develops.

If we're talking about a pregnancy in space, it's probably in the context of a longer mission, where the person's likely to be there for a longer period of time.

Plus, we've got to factor in the travel time—so to get to Mars, we're looking at almost a year. It's a lot of radiation if we add it all up. Plus, it's the kind of radiation that we have no idea how it would affect an unborn child. That's what makes it dangerous.

There's also a huge gravitational load on the body when you go into space and come back. The other impact is the weightlessness of the space environment; we already know that astronauts lose bone and muscle when they go into space. There's also an impact on the blood and heart system within the body. And even though the reproductive system doesn't change in space, it does need all those other systems to work. When women get pregnant, they need an effective blood supply so they can get nutrients to the baby. But in space, your overall blood volume is reduced. So would the baby get enough nutrients? There are so many unknowns [3].

How Realistic Are Plans to Start Extraterrestrial Human Colonies?

When it comes to space exploration, the reason that it's advanced so much is because of human curiosity. I think there's a lot of research that needs to be done to ensure that human beings who do travel to other planets are safe and that we're not putting anyone at risk.

Bibliography

1. Komorowski M, Fleming S, Kirkpatrick AW. Fundamentals of anesthesiology for spaceflight. J Cardiothorac Vasc Anesth. 2016;30:781.
2. Jennings RT, Baker ES. Gynecological and reproductive issues for women in space: a review. Obstet Gynecol Surv. 2000;55(2):109.
3. Drago-Ferrante R, Di Fiore R, et al. Extraterrestrial gynecology: could spaceflight increase the risk of developing cancer in female astronauts? An updated review. Int J Mol Sci. 2022;23(13):7465.

Forensic Medicine and Clinical Psychology in Space 10

We're working out how to solve crimes in space – the final frontier of forensic science.

Published: 7 March 2024

An intriguing new field emerges from the final frontier: astroforensics. This discipline is still in its infancy but is propelled by the inevitability of human nature. Space presents a unique and harsh environment for forensic investigations. Settings that involve altered gravity, cosmic radiation, extremes in temperature, and the need for oxygen-providing climate systems provide a few examples of the unearthly variables faced by future explorers.

The significant reduction of gravity in space introduces novel challenges in understanding this concept. We are at the beginning of a new research era, exploring the impact of the extraterrestrial environment on the behavior of forensic evidence. To extend research in this new forensic discipline, larger microgravity environments will be required, and the authors would be more than happy to operate the galaxy's first extraterrestrial forensic science laboratory.

If the crime scene moves into space, how to solve crimes in extraterrestrial environments? The latest frontier of forensic science seeks to understand how the unique environment of space will influence scientific crime investigations, for example, the analysis of bloodstains in microgravity [1].

The extreme environment, alteration of the night-day cycle, confinement in narrow spaces, lack of privacy, isolation, and distance from loved ones are the main stress factors to which astronauts are subjected when they spend long periods in space. (*Apr. 21, 2021*).

R. G. Russo, *Navigating Health Challenges in Space*,
https://doi.org/10.1007/978-3-032-03620-9_10

Clinical Psychology

Clinical psychology in space, or space psychology, is a specialized field that focuses on the mental health and well-being of astronauts in the space environment. Aerospace clinical psychological services usually are offered on a unit or organizational level, though interventions can be designed for individuals and their families. These services address the unique challenges of space travel, including isolation, confinement, and the physical effects of spaceflight, on astronaut mental health. Space psychologists work to prevent and treat mental disorders, improve teamwork and communication among crew members, and understand how space affects cognitive function and personality [2].

Bibliography

1. Alfano CA, Bower JL, et al. Mental health, physical symptoms and biomarkers of stress during prolonged exposure to Antarctica's extreme environment. Acta Astronautica. 2021;181. Arone A, Ivaldi T, et al. The burden of space exploration on the mental health of astronauts: a narrative review. Clin Neuropsychiatry. 2021;18(5): 237–46.
2. King R. Aerospace clinical psychology and its role in serving practitioners of hazardous activities. Hum Perf Extrem Environ. 1999;1:109–11.

General and Applied Hygiene 11

Astronautical hygiene works with space medicine to ensure astronauts are healthy and safe.

Astronautical hygiene evaluates and mitigates hazards and health risks to those working in low-gravity environments [1]. The discipline of astronautical hygiene includes topics such as the use and maintenance of life support systems, the risks of extravehicular activity, the risks of exposure to chemicals or radiation, the characterization of hazards, human factor issues, and the development of risk management strategies. Astronautical hygiene works side by side with space medicine to ensure that astronauts are healthy and safe when working in space.

Astronauts traveling in space are exposed to numerous hazards, such as radiation, microbes in the spacecraft, and planetary surface toxic dust [2]. During a space voyage, astronautical hygienists work on collecting data concerning a multitude of subjects. Once the data has been collected, they then analyze the data to determine, among other things, the risks to human health due to exposure to the various chemicals within the spacecraft as well as other toxins during their flight. From that, the hygienists can determine the appropriate measures to take to mitigate exposure of the astronauts to harmful chemicals.

The astronautical hygienist's main roles are as follows [3]:

(a) To initiate and participate in research where a competent assessment of the risks to health are critical, e.g., in the development of effective dust mitigation strategies for lunar exploration.
(b) To be actively involved in designing hazard mitigation techniques—for example, space suits with low dust retention/release and ease of movement.
(c) To provide in-flight troubleshooting, e.g., for identifying a hazard, assessing its health risks, and setting up measures for mitigation.
(d) To advise governments such as the UK Space Agency on the most cost-effective risk mitigation measures for crewed space flights.
(e) To act as a central link between other space science disciplines.

R. G. Russo, *Navigating Health Challenges in Space*,
https://doi.org/10.1007/978-3-032-03620-9_11

(f) To provide information, instruction, and training on standard setting, the health effects of exposure, hazard identification, and means of control.
(g) To devise a holistic approach to protecting an astronaut's health.

Personal Hygiene

Cleaning and waste disposal issues arise when dealing with low gravity environments. On the International Space Station, there are no showers, and astronauts instead take short sponge baths, with one cloth used to wash and another used to rinse. Since surface tension causes water and soap bubbles to adhere to the skin, very little water is needed [3].

Furthermore
The International Space Station (ISS) is a space station in low Earth orbit, dedicated to scientific research and managed by five different space agencies: the US NASA, the Russian RKA, the European ESA (with all related space agencies), the Japanese JAXA, and the Canadian CSA-ASC.

During their stay on board the ISS, astronauts must live and work in an environment very different from what they are used to on Earth. They must wash, go to the toilet, eat, drink, and keep fit and healthy in this environment [4].

Purposes

The International Space Station primarily serves as a scientific research laboratory, providing the advantage of being a long-term platform in a space environment, in which long-duration experiments can be conducted in weightlessness [5]. It also allows the experiments and components of the spacecraft itself to be monitored, integrated, repaired, and replaced.

The main research fields include human research, space medicine, biology (with biomedical and biotechnology experiments), physics (including fluid mechanics and quantum mechanics), materials science, astronomy (including cosmology), and meteorology.

Physical Exercise

The most negative effects of long-term weightlessness are muscle atrophy and spaceflight osteopenia. Other significant effects include fluid redistribution, slowing of the cardiovascular system, reduction in red blood cell production, balance disorders, and weakening of the immune system. Minor symptoms include loss of body mass, nasal congestion, sleep disturbances, excess flatulence, and facial swelling. These effects quickly disappear upon return to land. Astronauts use bungee cords to attach themselves to the treadmill. Researchers believe that exercise is good protection for bones and also helps limit the loss of muscle mass that occurs when living without gravity for a long time.

Bibliography

1. Orion Quick facts. NASA; 2014. Cain JR. Astronautical hygiene-A new discipline to protect the health of astronauts working in space. J Br Interplanet Soc. 2011;64:179–85. Personal Hygiene Provisions. NASA; 2002. Archived from the original on June 20, 2000. Retrieved September 5, 2007.
2. K. Space Hygiene. Space Biomedical Research Institute; 1998.
3. Personal Hygiene Provisions. NASA; 2002.
4. International Space Station facts and figures. nasa.gov. NASA; 2021.
5. Columbus laboratory. European Space Agency (ESA); 2009. Ferroni E. The risks of astronauts in space; 2014.

Part III

Viruses and Bacteria in Space

Microbiology and Infectious Diseases 12

Most bacteria are harmless, and many are beneficial or even essential to human functioning and well-being. But microgravity can make some microbes more likely to cause disease, and bacteria and fungi may affect the function of spacecraft systems, for example, by corroding metal on spacecrafts. These organisms also could contaminate other planetary bodies that spacecraft and humans land on [1].

Because of the potential for microbes to cause infection in astronauts and to degrade various components that may be vital for the functioning of the spacecraft, the risks must be assessed and, where appropriate, microbial growth must be controlled through good astronautical hygiene, for example, by frequently sampling the space cabin air and surfaces to detect early signs of a rise in microbial contamination, keeping surfaces clean by the use of disinfected cloths, ensuring that all equipment is well maintained in particular the life support systems, and regular vacuuming of the spacecraft to remove dust. Further research in this field is therefore needed so that the risks of exposure can be evaluated and the necessary measures to mitigate microbial growth are developed.

The absence of gravity, solar radiation, and suboptimal nutrition are among the main causes of various dysfunctions found in organs after space missions. Among the organs most affected are the eyes, heart, and even bones.

Microbiology applied to space is an emerging sector in space sciences, and it is in the interest also of ASI (Agenzia Spaziale Italiana, Italian Space Agency) to create a common platform to support national excellence, in order to create synergies and identify future opportunities [2].

In recent years, microbiologists and astronauts have been testing tools that allow DNA sequencing directly in orbit—this is an important step to make future journeys into deep space even safer and more efficient.

Wherever there are humans, there are microbes, too. Bacteria and fungi live all around us, in our homes, offices, industrial areas, the outdoors—and even in space. The trick is limiting potentially harmful ones, particularly in a contained

R. G. Russo, *Navigating Health Challenges in Space*,
https://doi.org/10.1007/978-3-032-03620-9_12

environment such as a spacecraft. So from the launch of the very first module of the International Space Station, NASA has monitored its microbial community [3].

It is every man's dream to 1 day travel between stars and planets, walk in space, and float in spaceships in zero gravity. Although in the collective imagination it is one of the most exciting experiences that can be had, leaving for a mission in space is not always "roses and flowers," on the contrary. Not all of us are aware of the risks faced by those who remain in orbit around the Earth for months and months: osteoporosis, space nausea, loss of bone and muscle mass, heart problems, space blindness, and diabetes—all of this caused by space microbes, solar storms, zero gravity, radiation, and toxic dust. It seems like a war bulletin from a science fiction author, but it is reality, exactly what the astronauts experience both during the mission and upon returning to Earth.

It is known that there are numerous physiological, as well as psychological, consequences that astronauts experience after a long stay in space. A long stay, as specified by the National Academy of Sciences in a recent report, also means a 30-day journey, even if, in most cases, astronauts remain in orbit for 3–12 months (even 18 at times).

The human body will have to get used to microgravity [4], which is 38% compared to ours: muscles and bones would be subjected to a lot of stress, not to mention solar radiation [4]. NASA is required by law to protect its crew from events such as solar storms, because even the cumulative exposure to radiation that occurs in space is a real risk for health [5] as it significantly increases the chances of developing tumors. NASA experts have estimated in the past that to stay below a 3% risk rate, a man would have to spend a maximum of 268 days in space, and a woman 159.

A space adaptation syndrome (SAS) [6] affects many astronauts: among the most unpleasant consequences is space nausea, similar to seasickness.

The long-term consequences of the changes could be very serious once they return to Earth and even lead to more serious heart problems. Two recent studies have also specifically examined the changes and stress that the eyes undergo in conditions of orbital microgravity: oxidative stress of the eyeballs, i.e., the rapid aging of the eye, was found. Spaceflights subject astronauts' eyes to harsh radiation, hypothermia, hypoxia, and changes in gravity, which cause tissue damage.

Wherever there are humans, there are microbes, too. Bacteria and fungi live all around us, in our homes, offices, industrial areas, the outdoors—and even in space. People literally could not live without these tiny organisms, many of which are beneficial.

The trick is limiting potentially harmful ones, particularly in a contained environment such as a spacecraft. So from the launch of the very first module of the International Space Station, NASA has monitored its microbial community.

Microbiological Hazards

During spaceflight, transfer of microbes between crew members [7] is common. The Skylab 1 crew was affected by several diseases of bacteriological origin. Microbial contamination in the Skylab was found to be very high. *Staphylococcus aureus* and *Aspergillus* have been detected from the air and on surfaces during several space missions. Microbes do not settle in microgravity, which results in persistent airborne aerosols with high microbial densities, particularly if air filtration systems are not well maintained.

Urine collection devices produce the bacterium *Proteus mirabilis*, which is associated with urinary tract infections, and for this reason, astronauts can be very exposed to this type of infection [1]. An example is the Apollo 13 mission, during which the lunar module [8] pilot contracted an acute urinary tract infection which was then treated with 2 weeks of antibiotics.

Biofilms can contain a mixture of bacteria and fungi that can damage electronic equipment by oxidizing various components, such as copper wires. These organisms flourish because they survive on organic matter released from the astronaut's skin. Organic acids produced by microbes, especially fungi, can corrode steel, glass, and plastic. Additionally, on a spacecraft, due to increased radiation exposure, there are likely to be more microbial mutations.

Given the possibility that microbes cause infections and have the potential to damage various components vital to the operation of the spacecraft, it is important that risks are assessed and, if necessary, managed by monitoring levels of microbial growth and by observing good astronautical hygiene.

For example, early signs of increased microbial contamination [9] can be detected by frequently sampling cabin air and surfaces, keeping surfaces clean with the use of disinfected cloths, ensuring all equipment is properly maintained, especially life support systems, and regularly vacuuming the spacecraft to remove dust.

Because of the potential for microbes to cause infection in the astronauts and to degrade various components that may be vital for the functioning of the spacecraft, the risks must be assessed and, where appropriate, the levels of microbial growth must be controlled by good astronautical hygiene.

Microbes and Microgravity in Space

More than a hundred strains of bacteria and fungi have been identified from manned space missions. These microorganisms survive and reproduce in space. Much effort is being made to significantly reduce the risks arising from microbial contamination. Spacecraft are routinely sterilized by washing with antimicrobial agents such as ethylene oxide and methyl chloride, and astronauts are quarantined for several days before a mission. However, these measures only succeed in reducing microbial populations, not eliminating them. Microgravity can boost the virulence of specific microbes. It is therefore important to study the mechanisms responsible for this and take all the necessary measures to make sure that astronauts, particularly those who may be immunocompromised, are not affected.

Antimicrobial Resistance (AMR) Virulence in Space [10]

Two common bacteria, *Enterococcus faecalis* and *Enterococcus faecium*, previously have been isolated on the International Space Station and show antibiotic resistance and tolerance to desiccation, starvation, and disinfection. Are bacteria more virulent and drug-resistant in space than they are on Earth? The relationship between microbial adaptations and host health in space is somewhat murky. One study showed that the microbial resistance potential of the ISS microbiome was similar to controls on Earth, suggesting the threat of AMR is comparable.

Viruses and Bacteria in Space

There are various biological threats in the cosmos. More than 250 different species of bacteria and fungi can survive in outer space. Bacteria always pose a significant contamination challenge in spacecraft assembly clean rooms despite decontamination procedures.

Studies suggest pathogens can become more infectious in space. High radiation levels may cause viruses to mutate more frequently. At the same time, the harsh conditions in space can weaken astronauts' immune systems, making them more susceptible to infections.

Viruses were found at relatively low levels, based on a metagenomic analysis. "Spaceflight sometimes reactivates viruses (e.g., herpes, Epstein-Barr, varicella-zoster, cytomegalovirus) and increases viral shedding in astronauts. A human infection in a space-like environment could pose a significant threat to everyone on board" [11].

A Few Virus Oddities in Space: Can Viruses Survive in Space?

Astronauts already infected when they blast off into space spread viruses to their fellow fliers, and viruses can easily jump from host to host in the earthlike environment of space shuttles or the International Space Station (ISS). In outer space, there are no hosts, and the harsh environment is a far cry from "the best possible conditions." That said, viruses do not last long in space at all—and almost certainly not long enough for a spacecraft to reach another world.

How Long Can a Virus Live Outside a Body?

Viruses can live for a surprisingly long time outside a body, depending on conditions such as moisture and temperature. However, they tend to live longer on water-resistant surfaces, such as stainless steel and plastics.

A common cold virus can sometimes survive on indoor surfaces for several days, although its ability to cause infection drops dramatically over time. Flu viruses can survive in the air for several hours, especially at low temperatures, and on hard surfaces, they can remain infectious for 24 hours.

Enteric viruses such as norovirus (pictured) and hepatitis A can survive for weeks on a surface if conditions are suitable. The norovirus is known for causing sickness outbreaks in schools, cruise ships, and hospitals.

Why Do Viruses Reactivate in Space?

It is believed that stressful life situations cause viral reactivation. NASA research published in Frontiers shows that these latent virus strains can reactivate in people who spend time in space.

What Viruses Were Detected in Astronauts?

There have been many cases of viruses being transported to space unintentionally by astronauts or spacecraft. The best known one is the Apollo 7 mission when the crew got very sick with rhinovirus.

How Do Dormant Viruses Behave During Space Flight?

The immune system controls them and keeps the virus in a "dormant" state in the spinal cord nerve cells. Viruses can, however, intermittently "awaken" or reactivate, starting to grow and multiply. Most times when this happens, the person experiences no symptoms or overt illness; however, the reactivation does cause disease more in elderly people and in people with weak immune systems. It is believed that stressful life situations cause the viral reactivation.

Dormant Viruses Can Reactivate in Anyone, Not Just Astronauts

To study the physiological impact of spaceflight, saliva, blood, and urine samples collected from astronauts before, during, and after spaceflight have been analyzed.

During spaceflight, there is a rise in the secretion of stress hormones like cortisol and adrenaline, which are known to change the immune system. In keeping with this, we find that astronauts' immune cells—particularly those that normally suppress and eliminate viruses—become less effective during spaceflight and sometimes for up to 60 days after they return home to Earth.

Four of the eight known human herpes viruses—including the variety responsible for chickenpox and shingles (VZV)—have been detected in astronauts during their stay in space. Herpes viruses often remain dormant in nerves or immune cells throughout a lifetime.

Herpes Virus Hazards [12]

Reactivation of the herpes virus has been documented in more than half of astronauts. The cause appears to be stress. This could create problems for future space missions.

Many people on the International Space Station have suffered from skin rashes and mild respiratory infections, and researchers have found that space travel can reactivate dormant viruses, such as herpes simplex, chickenpox, and cytomegalovirus.

Herpes Threatens Space Missions

NASA has issued a warning about space herpes after a study found the virus was reactivating in crew aboard the Space Shuttle and International Space Station missions. Reactivation of dormant herpes virus was documented in more than half of astronauts. The cause appears to be stress. Although only a small proportion of astronauts showed symptoms, this poses a serious problem that could have implications for future space missions.

The stress of spaceflight on the body is believed to contribute to suppressing the immune system and helping the virus grow. It has already been noted that during spaceflight, there is a rise in secretion of stress hormones like cortisol and adrenaline, which are known to suppress the immune system.

Do Astronauts Get Sick More Easily While in Space?

The genetic and physiological changes that occur in microgravity can impact the health of astronauts, even when they return to Earth. Traveling in space, in fact, causes enormous physiological changes. The most significant is the shift of body fluids from the lower body to the trunk and head, and as astronauts float in microgravity, they can suffer anemia, eye problems, changes in blood pressure, and immune dysfunction.

"Latent herpes virus reactivation has been demonstrated in astronauts during shuttle (10–16 days) and International Space Station ($\geq$180 days) flights. Following reactivation, viruses are shed in the body fluids of astronauts. Typically, shedding of viral DNA is asymptomatic in astronauts regardless of mission duration; however, in some cases, live/infectious virus was recovered by tissue culture that was associated with atopic-dermatitis or skin lesions during and after space flight. Four of the eight known human Herpes viruses – including the variety responsible for chickenpox and shingles (VZV) – were detected in astronauts during their stay in space. Herpes viruses often remain dormant in nerve or immune cells throughout a lifetime." "Immunosuppression, bacterial virulence, and the resulting risk of infection increase with the duration of the space mission. We must continue to develop new approaches to combat bacterial infections if we are to attempt longer missions to Mars and beyond."

Other Microbe Oddities in Space [13]

What Unknown Bacteria Were Found in Space?

Thirteen strains of *Enterobacter bugandensis*, a bacterium notorious for being multidrug resistant, have been isolated from the ISS and studied.

(By Andrea Centini, 24 April 2024)

"Bacteria on the International Space Station have transformed into something never seen before. An international research team has observed that bacteria isolated on the International Space Station (ISS) have developed unique genetic, metabolic and functional characteristics, never before seen on Earth. They are a risk because they are super resistant to drugs.

According to experts, this radical divergence is linked to the extreme environment of the orbiting laboratory, in which bacteria must face multiple challenges; among them microgravity, high levels of radiation and CO2.

The discovery is significant because the bacterial species involved in the study—called *Enterobacter bugandensis*—is notoriously resistant to antibiotics, a superbug classified as an opportunistic pathogen, that is, capable of triggering infections in 'favorable' conditions, for example in sick, immunosuppressed and immunocompromised people. Studies on these unique strains found on board the ISS can be very valuable for protecting the health of astronauts—for example by preparing ad hoc antibiotics—and future human colonies on the Moon, Mars and elsewhere in space.

An international research team led by scientists from NASA's Jet Propulsion Laboratory (JPL) based at the California Institute of Technology, discovered that bacteria on the International Space Station have evolved in a completely different way from those on Earth. (CALTECH) of Pasadena. They collaborated with colleagues from the Department of Biotechnology at the Indian Institute of Technology in Madras and the Center for Integrative Biology and Systems mEdicine (IBSE). The researchers, coordinated by Dr. Kasthuri Venkateswaran, analyzed the bacterial samples collected by astronauts in various rooms of the ISS using special swabs and wipes, subsequently inserted into sterile containers and sent back to Earth for laboratory analysis. In a previous investigation, researchers had identified five strains of Enterobacter bugandensis, particularly in the space station's toilet and gym.

It is no surprise that bacteria brought by humans into space can evolve in a totally different way than what is seen on Earth."

Moreover, the red-colored bacterium *Deinococcus radiodurans* [2] can resist 1.5 million rads of gamma radiation, about 3000 times the amount that would kill a human. The bacterium survives and reproduces in environments that would be lethal for any other organism, and it also resists high doses of ultraviolet radiation. It is incredible that it lived on a specially designed platform outside the pressurized module of the ISS. After a year, after radiation, extreme temperatures, and absence of gravity, these bacteria were brought back to Earth: they were covered in small protuberances (vesicles) and numerous repair mechanisms had been activated.

Potentially Harmful

The new strains are both genetically and functionally different from their original terrestrial selves. The odd conditions of space have somehow made them more resistant to antibiotics and other treatment. The microgravity, solar radiation, and high carbon dioxide levels that the bacteria experience in low orbit have forced them to evolve into a kind of super-bacteria.

The Bacteria on the International Space Station Are Mutating

Genome analysis reveals how microorganisms are adapting to life in orbit, also strengthening some pathogenic traits.

Wherever there are humans, there are also bacteria. This is how microorganisms that evolved with us on Earth have also "conquered" space and are adapting. This is confirmed by the research of scientists at NASA's Jet Propulsion Laboratory, who have been collecting and analyzing samples from the International Space Station (ISS) for about a decade to monitor changes and ensure that there are no threats to astronauts in orbit [14].

The researchers isolated many bacterial species from samples taken in different environments of the ISS, comparing the genetic material. Microbes are differentiating from their relatives on Earth, developing adaptations that help them survive in extreme conditions.

In their latest work, Kasthuri Venkateswaran and colleagues studied bacterial species recently discovered on the ISS, in particular *Microbacterium mcarthurae*, *Microbacterium meiriae*, *Paenibacillus vandeheii*, *Arthrobacter burdickii*, and *Leifsonia williamsii*. These microbes—the researchers argue—have developed characteristics common to each other, but different from those of the same bacteria on Earth. They have adapted to life in space by modifying some proteins to make them more functional in microgravity conditions, and their DNA repair system is more active to counteract the effects of increased exposure to radiation. Furthermore, scientists observed the presence of mobile genetic elements that improved their metabolism. Study findings indicate under stress, the ISS isolated strains were mutated and became genetically and functionally distinct compared to their Earth counterparts.

Are Mutant Space Bacteria a Threat?

What particularly arouses the attention of experts, however, is the fact that genetic traits linked to pathogenic potential have emerged in the bacterial species living on the ISS: some genes associated with virulence (such as those that help evade or damage the immune system) are more active. Furthermore, it seems that space microbes are able to form biofilms on the surfaces of the ISS—an ability that makes them more resistant to disinfectants and antibiotics.

Whether this represents a concrete risk for the health of astronauts to date is not yet clear (there is however no alarm), but it is confirmed that studying the adaptations of microbes to space is of extreme importance in anticipation of long-term space travel. First of all, we are now aware that it is necessary to implement more incisive actions to prevent the proliferation of bacteria on surfaces (for example, better control of the humidity of the environments on the ISS); then, those same mutations could become targets for new drugs, if the modified bacteria prove to be a health risk.

It turned out that over 250 different species of bacteria and fungi can survive in outer space.

Bacteria pose a significant contamination challenge in spacecraft assembly clean rooms despite decontamination procedures. However, viruses were found to be present at relatively low levels, based on a metagenomic analysis. Studies suggest pathogens can become more infectious in space. Increased radiation levels may cause viruses to mutate more frequently. At the same time, the harsh conditions in space can weaken astronauts' immune systems, making them more susceptible to infections.

Viruses were found to be present at relatively low levels, based on a metagenomic analysis. "Spaceflight sometimes reactivates viruses (e.g., herpes, Epstein-Barr, varicella-zoster, cytomegalovirus) and increases viral shedding in astronauts. A human infection in a space-like environment could pose a significant threat to everyone on board" it notes.

Fungi in Space [15]

When fungi reproduce, they release spores into the air, sometimes forcefully! With no friction in space due to the lack of gravity, the spores simply float away. This could be why astronauts aboard Mir witnessed large, fungal formations seemingly all over the spacecraft. Fungi have a natural ability to survive and grow in extreme conditions. Fungi are one of the first organisms to thrive in environments on Earth that experience nuclear activity and, therefore, high levels of radiation.

Mushrooms are also relatively easy to grow and do not take much time, space, or resources to do so. It is for these reasons they are being considered as a sustainable food source for space travel. Species under consideration include oyster mushrooms on account of their fast growth and high yield.

Living spores have been collected in every level of Earth's atmosphere. Mushroom spores are electron-dense and can survive in a vacuum. Additionally, their outer layer is actually metallic and purplish colored, which naturally enables the spores to deflect ultraviolet light.

What Other Organs Are Involved?

The skeleton: Due to the absence of gravity as it is not necessary to counteract it, our body and muscles are inactive and calcium, instead of being deposited in the bones, is eliminated from the urinary system. Bones become thinner (osteoporosis), increasing the risk of developing kidney stones. Obviously, inactivity also affects the muscles, which tend to lose mass.

Other complications can occur in the airways, due to the accumulation of fluids in the upper part of the body which can cause severe nasal congestion and general breathing difficulty.

If the lucky astronaut did not experience problems with his bones, heart, and eyesight, space microbes would still undermine his health. A test performed on the Russian space station discovered the presence of 234 species of microscopic bacteria and fungi living on board with the astronauts. ISS personnel on duty between 1995 and 1998 reported a high number of microbial infections, such as conjunctivitis, acute respiratory distress, and dental infections. Medical studies have also proven that antibiotics are less effective, and for this reason, experts hope that antioxidants, which are taken with food and dietary supplements, can counteract these effects. You will gain a better understanding of the most common cardiovascular diseases, such as ischemic heart disease, heart valve dysfunction, glaucoma, and tumors.

Psychologists have proven that in space, astronauts experience fatigue, lethargy, fear of having appendicitis, tooth pain that appeared after dreaming of having tooth pain, and fear of becoming impotent. There are also positive effects, such as the total vision effect, that is, a sense of wonder and awe towards the universe, which leads to a feeling of peace and unity with nature and of transcendence.

Hazards Due to Lunar Dust

Lunar dust (or *regolith*) is the layer of particles on the surface of the Moon and is usually less than 100 μm [24]; the grain shapes tend to be elongated. Inhalation exposure to this dust can cause breathing difficulties due to the toxicity of the dust, it can also cloud astronauts' visors. It also adheres to space suits both mechanically (due to the grain shapes) and electrostatically. During the Apollo missions, dust also wore down the fabric of space suits.

Required measurements may include measurement of surface dust concentrations, surface electric fields, dust mass, velocity, charge and plasma characteristics. Its mechanical irritancy and chemical reactivity may also pose serious health risks to humans by a number of mechanisms.

Deposition of Inhaled Particles

The degree of inflammatory response in the lung depends on where the moon dust particles settle. Fine dust particles will settle in the lung alveoli, increasing the chance of lung damage.

Hazards from radiation, trapped radiation, cosmic radiation, solar particles and others: Thanks to the CRISPR-Cas9 system, it is possible to repair DNA damaged by the ionizing radiation to which astronauts on missions are exposed for long periods. This system is aimed at repairing DNA damaged not only by ionizing radiation but also by tumors.

CRISPR Cas9: This is the name of the genome editing system that earned scientists Emmanuelle Charpentier and Jennifer A. Doudna the 2020 Nobel Prize for Chemistry.

Space Medicine Conclusion

"Until 2025, space medicine will have contributed towards perceiving a human being as an integrative system, considering the complex interplay of all aspects of the process of ageing, and effectively developing measures of prevention, health preservation, and rehabilitation for patients on the ground as well as for astronauts on long-term missions."

From this point of view, the research that accompanies space missions certainly provides advantages such as the creation and development of new projects in the medical and clinical field. Space missions have also led to the development of advanced materials, which have then found numerous applications in modern medicine, for example, in the development of prosthetics such as artificial limbs, synthetic bones, and dental implants.

Bibliography

1. Hodkinson PD, Anderton RA, Posslt BN, Fong KJ. An overvview of space. Medicine. 2017;S1:i143–53.
2. Krittanawong C, Singh NK, et al. Human health during space travel: state of the art. Review. Cells. 2023;12(1):40.
3. Shirah B, Bukhari H, et al. Benefits of space medicine research for healthcare on earth. Cureus. 2023;15(5):e39174.

4. Willey JS, Britten RA, Blaber E, et al. The individual and combined effects of spaceflight radiation and microgravity on biologic systems and functional outcomes. J Environ Sci Health C Toxicol Carcinog. 2021;39(2):129–79.
5. Tran QD, Tran V, et al. Spaces medicines for space health. ACS Med Chem Lett. 2022;13(8):1231–47.
6. Quine T. Addicted to space: an appreciation of Anousheh Ansari, Part II. Spaceflight. 2007;49(4):144.
7. Hatyhre S, Wood JM, et al. Assessing the risk of transfer of microorganisms at the International Space Station Due to Cargo Delivery by Commercial Resupply Vehicles. Front Microbiol Sec Extreme Microbiol. 2020;11
8. Pohlen M, Carrol D, et al. Overview of lunar dust toxicity risk. Microgravity. 2022;8:55.
9. Chatterjee A, Abraham J. Microbial contamination, prevention, and early detection in food industry. In: Handbook of food bioengineering; 2018. p. 21–47.
10. Wazir HU, Narang P, et al. Bacterial virulence and prevention for human spaceflight. Life. 2023;13(3):656. Taylor PW. Impact of space flight on bacterial virulence and antibiotic susceptibility. Infect Drug Resist. 2015;8:249–62
11. Pavletić B, Runzheimer K, et al. Spaceflight virology: what do we know about viral threats in the spaceflight environment? Astrobiology. 2022;22(2):210–24.
12. Rooney BV, Crucian BE, et al. Herpes virus reactivation in astronauts during spaceflight and its application on earth. Front Microbiol. 2019;10
13. Makarova KS, Aravind L, et al. Genome of the extremely radiation-resistant bacterium Deinococcus radiodurans viewed from the perspective of comparative genomics. Microbiol Mol Biol Rev. 2001;65(1):44–79. Cowing K. Bacteria found on the international space station are mutating to become functionally distinct. Astrobiology. 2024; Woodford J. Bacteria on the space station are evolving for life in space. New Scientist. 2024; Sengupta P, Kerthick S, et al. Genomic, functional, and metabolic enhancements n multidrug-resistant *Enterobacter bugadensis* facilitating its persistence and succession in the International Space Station. Microbiome. 2024;12:62
14. Cortesão M, Schütze, et al. Fungal biotechnology in space: why and how? In: Nevalainen H, editor. Grand challenges fungal biotechnology. Grand challenges in biology and biotechnology. Cham: Springer; 2020.

Part IV

Pathology in Space

As humanity continues to push toward interplanetary travel and beyond, short- and long-term adaptations of the human body to extreme environmental conditions in space remain a fundamental concern. An astronaut is subjected to physically and mentally enduring situations like microgravity, extreme heat or cold, radiation, isolation, and other unforeseen technical challenges [1].

Bibliography

1. Wiencek JR, Das S, et al. Lab medicine in space. Clin Chem. 2023;69(5):442–9.

13 Alterations of the Cardiovascular System

When astronauts travel to space, their bodies begin to adjust to the microgravity of their spacecraft. Blood and other bodily fluids previously pulled downward by gravity now move toward the head, so the cardiovascular system does not have to work as hard to maintain blood flow to the brain.

Blood circulation too is obviously affected in space. The heart is a pump and a muscle at the same time: muscle contractions push blood through the body, and this circulation (which is influenced by gravity) ensures that the pump always has a blood supply to move.

Some alterations of the cardiovascular system have been identified: Orbital travel has even more specific consequences on the cardiovascular system: for example, the plasma loses volume, causing a possible increase in heart rate, and in general the blood in conditions of microgravity carries less oxygen.

Plasma, the liquid part of the blood loses volume, causing a possible increase in heart rate, and in general the blood in microgravity conditions carries less oxygen.

So How Does Blood Behave in Space?

In the absence of gravitational influence, the surface tension and cohesion of blood droplets are amplified. This means that blood in space has a greater tendency to adhere to surfaces until a greater force causes it to detach.

In spaceships and on the orbiting space station, the position of the body changes with respect to the center of gravity, which no longer exists, and therefore, the mass of blood fluid is redistributed in a different way than on Earth. Space travel imposes significant gravitational and radiation stress on both cellular and systemic physiology, resulting in a myriad of cardiovascular changes that have not been fully characterized.

It is known that in space, transmural cardiac pressure is increased as the normal gravitational compressive effect on the myocardium is withdrawn, resulting in

R. G. Russo, *Navigating Health Challenges in Space*,
https://doi.org/10.1007/978-3-032-03620-9_13

increases in stroke volume and cardiac output and lower left ventricular end-diastolic volumes.

Space cardiology is a rapidly growing field concerned with the cardiovascular adaptations to microgravity conditions and space stressors. Space travel results in significant gravitational and radiation stress on both the cellular and systemic levels, resulting in a myriad of physiologic changes that involve multiple organ systems (e.g., altered circulatory physiology, bone demineralization, optic disc edema).

The absence of orthostatic pressure and a decrease in arterial pressures reduce the workload of the heart and are believed to be the underlying mechanisms for the development of cardiac atrophy in space.

During spaceflight, astronauts are subjected to a prolonged state of microgravity (μg) that induces a myriad of physiological adaptations such as bone loss, muscle atrophy, fluid shifts, and decreased plasma volume as well as cardiovascular deconditioning, leading to orthostatic intolerance.

In space in microgravity, there is a loss of muscle mass and loss of bone matrix [3].

This loss is caused by the lack of body statism, which is an automatic reflex that keeps man standing on Earth. The main sensors of this reflex are in the soles of the feet and in the legs. Therefore, when gravity is lacking, the muscles no longer have the stimulus to contract to maintain balance; being inactive, they atrophy.

A mechanism of action similar to that which acts on muscles also occurs for bones, and in microgravity, a hormonal mechanism controlled by calcitonin is implemented, so that the bone becomes rarefied, undergoing osteoporosis.

To counteract muscle reabsorption, astronauts are encouraged to do a lot of physical exercise. The bone is a complex structure formed by cells that produce bone, osteoblasts, and cells that destroy it, osteoclasts. These cells are supported on a trabecular structure based on calcium, which is the bone matrix.

Space osteoporosis affects all astronauts, wasting 30–70% of bone, depending on the length of time spent in space, at an average rate of reduction of about 1.2% of mass per month [4]. It is a massive and worrying phenomenon that has affected all astronauts who have remained in space for periods longer than 3 months. There are also important cellular and molecular changes. Exposure to μg is associated with changes in cell shape and endothelial dysfunction through suppressed cellular proliferation, increased cell apoptosis, and oxidative stress.

Jugular Monitoring

The (internal) jugular vein, together with the other jugulars, has the task of carrying blood—now devoid of a good part of its oxygen—from the brain to the heart. This is essential to ensure that the circulatory system brings oxygenated blood to neurons, the cells of the brain, and removes cell waste and carbon dioxide.

As the researchers explain in their study, in two cases the astronauts' blood began to flow in reverse, from the heart to the brain. The hypothesis is that the inversion occurred due to a thrombus that prevented the normal venous flow towards the heart, a situation that can also occur on Earth in particular conditions of thrombosis.

NASA doctors then intervened: the astronaut had no particular symptoms, but as a precaution he was prescribed anticoagulant drugs to take for the rest of his stay in orbit.

All the astronauts' conditions gradually improved once they returned to Earth and normal gravity, returning to their pre-departure state. According to the researchers, this is a further sign that it is the space environment that conditions the blood supply and contributes to the formation of clots.

Deep vein thrombosis of the internal jugular vein of the neck was first detected in 2020 in an astronaut on a long stay on the ISS, requiring treatment with blood thinners. A subsequent study of 11 astronauts found slowed blood flow in the neck veins and even reversal of blood flow in two of them. NASA is currently conducting more research to see whether these abnormalities predispose astronauts to blood clots.

The (internal) jugular vein, together with the other jugulars, has the task of carrying the blood—now devoid of a good part of its oxygen—from the brain to the heart. This is essential to ensure that the circulatory system brings oxygenated blood to neurons, the cells of the brain, and removes cell waste and carbon dioxide.

As the researchers explain in their study, in two cases the astronauts' blood began to flow in reverse, from the heart to the brain. The hypothesis is that the inversion occurred due to a thrombus that prevented the normal venous flow towards the heart, a situation that can also occur on Earth in particular conditions of thrombosis. NASA doctors then intervened: the astronaut had no particular symptoms, but as a precaution he was prescribed anticoagulant drugs to take for the rest of his stay in orbit.

Cardiac Rhythm

Among astronauts, most of the heart disorders have been related to the cardiovascular system, and it is not clear whether they reflected existing conditions or effects of spaceflight. It is hoped that advance screening for coronary disease will greatly mitigate this risk. Other heart rhythm problems, such as atrial fibrillation, can develop over time, calling for periodic screening of crew members' heart rhythms. Beyond these terrestrial heart risks, some concern exists that prolonged exposure to microgravity may cause heart rhythm disturbances. Although this has not been observed to date, further surveillance is warranted.

The astronauts' hearts at risk have been widely studied and discussed. Below are some observations:

In the light of its research findings, NASA is evaluating the launch of a program to monitor astronauts in orbit, with periodic analyses to detect risks associated with thrombosis. Flight doctors have also established that anticoagulant drugs and other medicines must always be available on the ISS to treat any problems.

Conclusion

The main changes in the cardiovascular system induced by weightlessness involved blood pressure, central venous pressure, baroreflexes, cardiac output, and systolic output. Our understanding of some aspects is still not clear, such as postflight orthostatic intolerance and the more recent SANS ("neuro-ocular syndrome associated with spaceflight"). The SANS syndrome, according to some scholars, presents analogies with the terrestrial syndrome of "idiopathic intracranial hypertension."

Numerous countermeasures are available to astronauts to mitigate the effects on the various systems of the human body. During weightlessness, austronauts perform daily aerobic, dynamic, and resistive training according to protocols tested on the ISS to counteract muscle and bone breakdown and the decrease in aerobic capacity. The exercises are integrated with a diet that includes antioxidants, omega-3 fatty acids, and small amounts of salt, iron, and bisphosphonates.

Finally, data on the current population of astronauts support the notion that the primary cardiovascular risks of spaceflight are compromised hemodynamic responses to central hypovolemia, resulting in reduced orthostatic tolerance and exercise capacity rather than actual cardiac dysrhythmias, reduced cardiac contractile and vascular function, or any asymptomatic cardiovascular disorder. Environmental space hazards act on the cardiovascular system through multiple mechanisms. Microgravity causes cardiac atrophy and orthostatic intolerance, while space radiation can potentially accelerate atherosclerosis. Further research is needed, especially regarding long-term spaceflights.

Definitions: Alterations of the Cardiovascular System

Term	Definition
Aneurysm	Localized dilation of a vessel wall or cardiac chamber Caused by weakening of vessel wall
Embolism	Detachment of a thrombus
True and false aneurism	A true aneurysm is one that involves all three layers of the wall of an artery (intima, media, adventitia) A false aneurysm (pseudoaneurysm) is a collection of blood leaking completely out off an artery or vein, but confined next to the vessel by the surrounding tissue
Hypertension	Elevated blood pressure above normal range
Pericardial effusion	Accumulation of fluid in the pericardial cavity Transudate: Serous/clear fluid Exudate: Infected fluid
Regurgitation	"Leaky" valves lead to backflow of blood For example, mitral regurgitation, would increase workload (preload) on the left ventricle and decrease ejection of blood into the aorta and body circulation
Stenosis	Narrowing of the valves; restricts blood flow through the heart

Term	Definition
Thromboembolism	A condition in which thrombosis develops followed by embolism, e.g., DVT followed by PE
Thrombus	Blood clot inside blood vessels

Disorders of the Cardiovascular System [5, 6]

Arteriosclerosis

Stiffness, loss of elasticity, and thickening of the arterial wall
Leads to narrowing and subsequent obstruction of the lumen, resulting in decreased blood flow

Atherosclerosis:

- Most common form of arteriosclerosis
- Can affect any vessel
- A chronic inflammatory response to lipid accumulation in the intimal layer of the arterial walls
  - Over time, there is progressive hardening of the artery, loss of elasticity, and plaque formation (atheromas)
  - Narrowing of the artery leads to low blood supply
  - The process starts from birth and continues throughout the lifespan
  - Lifestyle (i.e., diet, activity, substance use) impacts the rate and severity of plaque build-up

Manifestations

When atherosclerosis develops in a coronary artery, this leads to coronary artery disease (CAD), which can progress to angina and MI.

Non-modifiable Risk Factors

- Age
- Gender
- Family history

Modifiable Risk Factors

- Lifestyle
- Obesity
- Physical inactivity
- Smoking

Diseases That Increase Risk for Atherosclerosis

- Type 2 DM (can be reversed)
- Those with poor diabetes management are at increased risk.
- Hyperlipidemia
- Hypertension

Atherosclerosis Pathophysiology:

- Endothelial damage
- Multiple causes for endothelial cell damage (e.g., smoking, HTN, diabetes, high LDL, and inactivity)
- Damaged endothelium loses its protective antithrombic vasodilating mechanisms
- Damaged endothelium evokes inflammation with attraction of inflammatory cells including macrophages and generation of oxygen free radicals
- Oxidized LDL in the intima recruits monocytes that differentiate into macrophages

Fatty streaks/plaque

- Lots of foam cells create what is known as a fatty streak
- Fatty streaks produce more free radicals and inflammatory mediators
- Macrophages also release growth factors that stimulate smooth muscle proliferation

Fibrous plaque

- Collagen grows over the streak to make a fibrous cap fibrous plaque
- Can calcify and then protrude into the lumen to obstruct blood flow

Complicated plaque and thrombus formation

- Some plaques can become "unstable" and can rupture
- A plaque that ruptures is a "complicated plaque"
- Exposure of the underlying tissue
  - Results in platelet adhesion, initiation of the clotting cascade, and thrombus formation
- Thrombus can occlude a vessel
  - Partial occlusion leads to ischemia
  - Complete occlusion leads to infarction

Summary of atherosclerosis pathophysiology:

- Stage 1: Scratch and damage to the endothelium

- Stage 2: Inflammation, collection of macrophages, and other inflammatory cells
- Stage 3: This makes it easy to generate a thrombus on top of plaque or exposed suburothelium

Coronary Artery Disease (CAD)

- Atherosclerosis of the coronary arteries
- Narrowing or blocking of coronary arteries → decreased blood flow → low O_2 supply to the heart → low tissue perfusion→ myocardial ischemia

Clinical manifestations:

- Asymptomatic
- Angina pectoris
- Myocardial infarction
- Cardiac complications
- Dysrhythmia
- Heart failure
- Sudden death

Bibliograhy

1. Wiencek JR, Das S, et al. Lab medicine in space. Clin Chem. 2023;69(5):442–9.
2. Baran R, Marchal S, et al. The cardiovascular system in space: focus on in vivo and in vitro studies. Biomedicines. 2021;10(1):59. Ed. Arnab Ghosh
3. Baran R, Marchal S, et al. The cardiovascular system in space: focus on in vivo and in vitro studies. Biomedicines, Editor: Arnab Ghosh. 2021;10(1):59.
4. Shykoff BE, Farhi LE, et al. Cardiovascular response to submaximal exercise in sustained microgravity. J Appl Physiol. 1996;81:26–32.
5. Vernice NA, Meydan C, Afshinnekoo E, Mason CE. Long-term spaceflight and the cardiovascular system. Precis Clin Med. 2020;3:284–91.
6. Grimm D, Wehland M, Corydon TJ, Richter P, Prasad B, Bauer J, Egli M, Kopp S, Lebert M, Krüger M. The effects of microgravity on differentiation and cell growth in stem cells and cancer stem cells. Stem Cells Transl Med. 2020;9:882–94.

Alterations of the Musculoskeletal System in Space

14

The musculoskeletal system, which includes the bones, ligaments, tendons, and cartilage, is an essential component of human health. In addition to providing the body with structure and the means for movement, the musculoskeletal system acts as an endocrine system, stimulated by exercise, interacting through biochemical signaling with other organs in the body. The main sensors are in the soles of the feet and legs; therefore, with the lack of gravity, the muscles no longer have the stimulus to contract to maintain balance; being inactive, they atrophy.

Exercise, working through muscle and bone interactions, plays an important role in the maintenance of health. Lack of exercise leads to obesity and increased complications such as diabetes and metabolic disease.

As people age, they face new challenges in maintaining musculoskeletal health, as the interactions between muscle and bone are complicated by the aging process. As the population ages, the conditions of sarcopenia, osteoporosis, and arthritis will increase as will hip and knee replacements.

Musculoskeletal disorders are the leading cause of disability in the United States, with an enormous impact on quality of life and longevity. They encompass a broad range of conditions that impair normal activity due to injury, pain, or disease of the bones, joints, ligaments, muscles, and tendons. Many of these disorders are associated with, or made more complex by, the aging process. The IU School of Medicine has developed a comprehensive team of researchers and clinicians working to better understand the science and improve care and treatment of these disorders.

The skeletal system has several jobs, some of which are easier to see and feel than others. But they are all important to your body and health. Your skeletal system's major functions include:

- Giving your body shape: Your skeleton supports the weight of your body. It is also the anchor for all the tissue that connects to it.
- Helping you move: Your joints, connective tissue, and muscles all work together to push and pull parts of your body every time you move.

R. G. Russo, *Navigating Health Challenges in Space*,
https://doi.org/10.1007/978-3-032-03620-9_14

- Protecting your organs: The human skeleton is like a built-in suit of armor for organs throughout your body. Your skull protects your brain, your ribs shield your heart and lungs, and the vertebrae in your spine keep your spinal cord safe.
- Creating new blood cells: The bone marrow at the center of some of your bones produces new red blood cells and white blood cells your body needs to stay healthy.
- Storing minerals: Your bones hold your body's supply of important minerals like calcium and vitamin D.

Microgravity Effects on Muscle

Skeletal muscles perform several functions that are essential for human life, as they not only provide the strength needed to counteract gravity and maintain posture, but also constitute a storehouse of important substrates, such as amino acids and carbohydrates, that are crucial for heat and energy production [1].

A hormonal mechanism controlled by calcitonin is implemented, whereby the bone becomes rarefied, undergoing osteoporosis.

To counteract muscle resorption, astronauts are advised to do plenty of exercise. Bone is a complex structure made up of cells that produce bone, osteoblasts, and cells that destroy it, osteoclasts. These cells are supported on a calcium-based trabecular structure, which is the bone matrix.

Space osteoporosis affects all astronauts, dissipating 30–70% of the bone, depending on the length of time spent in space, at an average rate of reduction of about 1.2% of mass per month. It is a massive and worrying phenomenon that affected all astronauts who remained in space for periods longer than 3 months.

What Happens to the Skeleton in Space?

In the microgravity of space, astronauts lose an average of 1–2% of their bone mineral density each month. This loss occurs in the weightless environment of space because the bones no longer have to hold the body against the pull of gravity. The musculoskeletal system enables the human body to move and exert forces in its environment. While on Earth this system must support the weight of the body and the bones are always under load due to gravity, in the absence of gravity and there is "no weight," this does not happen. When astronauts are in orbit, the weight the body must support is almost nil, so the musculoskeletal system is not loaded and no action is tiring. The consequence of staying in conditions of microgravity (the total absence of gravity or zero gravity is only a theoretical situation) is therefore the loss of bone and muscle tissue.

With bone structure remodeling and/or bone loss during spaceflight at the rate of 1–2% per month, after 6 months in space, astronauts present osteoporosis symptoms similar to those seen in older women on Earth. Over time, in fact, the calcium that normally accumulates in bones is mobilized and released into the blood. The amount of calcium in the blood of astronauts during space missions is much greater than it would be on Earth and reflects the loss of bone density. This decrease, known as disuse osteoporosis (similar to that seen on Earth in the case of prolonged immobility), weakens the bones and makes them less able to support body weight and movement once returned to Earth.

In microgravity conditions, load-bearing bones such as the spine, pelvis, and lower limbs are those most affected, and up to 20% bone mass loss has been reported after a six-month mission. Astronauts returning from a long space mission therefore risk fractures and are consequently subjected to specific attention and care. Astronauts restore most of their bone mass in the months following their return from space, even if they do not fully recover.

Loss of Bone Density

Spaceflight osteopenia is the bone loss associated with human spaceflight. The metabolism of calcium is limited in microgravity and will cause calcium to leak out of bones. After a 3–4-month trip into space, it takes about 2–3 years to regain lost bone density. New techniques are being developed to help astronauts recover faster. Research in the following areas holds the potential to aid the process of growing new bone:

- Diet and exercise changes may reduce osteoporosis.
- Vibration therapy may stimulate bone growth.
- Medication could trigger the body to produce more of the protein responsible for bone growth and formation.

The exact mechanism that causes calcium loss in microgravity is still not clear. Many scientists believe that microgravity increases the rate at which bone tissue is destroyed compared to that at which it is laid down. However, it is still not clear what triggers this difference in speed. Research is working on multiple fronts (hormone levels, diet, exercise) to determine the exact cause of osteoporosis during spaceflights so as to control or prevent it. In fact, the astronauts' osteoporosis is one of the major obstacles to the progress of space programs, since it would be extremely dangerous to expose them to very long stays in microgravity. To get to Mars and back, for example, currently it would take about 16 months and that is long enough to cause serious damage to the astronauts' bone and joint systems.

Muscle atrophy results from the lack of physically demanding activity (walking, climbing stairs, lifting objects, etc.) and from the lack of load on postural muscles. In short missions, muscle loss of 10–20% has been observed, and if no countermeasures are taken, this might reach 50% in long missions. This could compromise the astronauts' ability to do certain specific work requiring physical effort during the mission (for example, extravehicular activities) and after landing (for example, egress in emergency conditions). All this also involves discomfort: astronauts have often complained of pain in the lower back, which seems to be a consequence of atrophy of the postural muscles.

Loss of Muscle Mass

In space, muscles in the legs, back, spine, and heart weaken and waste because they are no longer needed to overcome gravity, just as people lose muscle when they age as a result of reduced physical activity. Astronauts rely on research in current areas to build muscle and maintain body mass: exercise can build muscle if at least 2 hours a day are spent doing resistance training routines. Neuromuscular electrical stimulation is one method to prevent muscle atrophy.

- Exercise may build muscle if at least 2 hours a day is spent doing resistance training routines.

Musculoskeletal Disorders in Space

Musculoskeletal disorders are a group of conditions that can affect the joints, muscles, and bones. There are over 150 musculoskeletal disorders, which include conditions such as back pain, arthritis, osteoporosis, and tendonitis. Symptoms of musculoskeletal disorders may include pain and a loss of mobility [2].

Bibliography

1. Tanaka K, Nishimura N, Kawai Y. Adaptation to microgravity, deconditioning, and countermeasures. J Physiol Sci. 2017;67:271–28.
2. Frontera WR, Ochala J. Skeletal muscle: a brief review of structure and function. Calcif Tissue Int. 2015;96:183–95. Narici MV, de Boer MD. Disuse of the musculo-skeletal system in space and on earth. Eur J Appl Physiol. 2011;111:403–20

15 Alterations of the Respiratory System

The respiratory system takes up oxygen from the air we breathe and expels the unwanted carbon dioxide. The main organ of the respiratory system is the lungs. Other respiratory organs include the nose, trachea, and breathing muscles (the diaphragm and the intercostal muscles). As the lungs receive approximately 100% of the cardiac output, the initial increase in astronauts increases pulmonary blood flow. Additionally, the removal of gravity allows for more uniform alveolar expansion, resulting in a greater surface area exposed to the external environment [1].

"The structure of the lung, with its delicate network of airspaces and capillaries, means that gravity has a profound influence on its function. Gravity serves to maintain a degree of matching of both ventilation and perfusion, so that the ventilation/perfusion ratio, and thus gas exchange, remains efficient."

In addition to the pulmonary congestion, which in the first days of stay causes a respiratory disorder, with a feeling of suffocation in some astronauts, the increase in liquid in the head causes congestion at the nasal level, with partial obstruction and coryza (a cold, in this case without bacterial or viral infection). This congestion is annoying and very accentuated for some, while it is almost non-existent for others. It heals upon return, almost instantly upon reentry to normal gravity [2].

What Changes Can Be Expected of the Respiratory System While in a Microgravity Environment?

As noted above, no change is observed in lung volume and capacity between G0 and 1G in long-duration spaceflights. There are either no or only slight changes in lung volume and capacity between G0 and 1G in parabolic and short-duration spaceflights. Pulmonary function is largely altered in microgravity, but none of the changes observed so far appear likely to limit long-term spaceflights.

R. G. Russo, *Navigating Health Challenges in Space*,
https://doi.org/10.1007/978-3-032-03620-9_15

The Breathing System in Space

Many of the tests are on the astronauts themselves, to determine the impact of life in space on the human body. Some are designed to utilize the uniqueness of the onboard lab's microgravity on non-space experiments such as crystal growth. But others are integral to space travel itself: exploring new technologies for use on the ISS and future longer manned space missions, to Mars, for example [3].

How Do Astronauts Breathe?

The short answer is that the astronauts bring oxygen from Earth and make oxygen by running electricity through water (this is called electrolysis). The air and water on the space station all originally came from Earth.

Breathing Fresh Air

We electrolyze water to split it open to hydrogen and oxygen. The oxygen goes into the atmosphere for the crew to breathe. The hydrogen is then combined with another waste product—carbon dioxide breathed out by the crew—to make methane and water.

The water gets reused and the methane is currently vented overboard. The NASA team is also developing a way to crack open methane to produce more hydrogen. If we had more hydrogen, we could be more efficient and produce more water, which we can then turn into more oxygen. So we would have to resupply less water for that process.

Studies of lung function in the absence of gravity provide valuable insight into how, for us Earth-bound individuals, its unavoidable effects shape our lung function. Gravity causes uneven ventilation in the lung through the deformation of lung tissue. Both ventilation and perfusion exhibit persisting heterogeneity in microgravity, indicating other important mechanisms. Therefore, while both ventilation and perfusion are more uniform in spaceflight, gas exchange is seemingly no more efficient than on Earth. Despite the changes in lung function when gravity is removed, the lung continues to function well in weightlessness. Unlike many other organ systems, the lung does not appear to undergo structural adaptive changes when gravity is removed, and so there is no apparent degradation in lung function upon return to earth, even after 6 months in space.

The Structure of the Lung Means That Gravity Has a Profound Influence on Its Function

The normal lung is markedly affected by the ever-present gravity we all experience, and these effects directly influence virtually every measurement of lung function that we make. For that reason alone, understanding how gravity affects the lung is a relevant topic, and just as we might place someone in a hot or cold environment in order to study thermoregulation, altering the level of gravity the lung experiences provides a means to study gravitational effects.

External Pressure

Perhaps, the most insidious danger in space is decompression. In the absence of adequate pressurization, the low pressure of space would cause the gases in the human body to expand. This could cause explosive trauma, with devastating consequences for tissues, organs, and even bone structure.

Bibliography

1. Barry M, Sumera K. Respiratory system. J Paramed Pract. 2020;12(8):1–4.
2. Peate I. The respiratory system. Br J Healthcare Assist. 2021;15(4):174–9.
3. Smith MB, Chen H, Oliver BGG. The lungs in space: a review of current knowledge and methodologies. Cells. 2024;13(13):1154.

16 Alterations of the Erythropoietic and Solar Eruptions

Erythropoiesis begins in the embryo during the third week of gestation, at the level of the yolk sac. From the fourth week, erythropoiesis begins in the liver, which becomes predominant at the end of the second month. From the fifth month, it begins at the level of the bone marrow, a site that becomes predominant at the end of gestation; this site remains the only one in the adult [1].

Researchers are studying which parts of an astronaut's body are most sensitive to solar radiation. Astronauts inside a space facility appear to be quite shielded from radiation emitted by the Sun, especially during solar flares, but astronauts protected by a space suit alone are very exposed to radiation from the Sun during a solar flare [2].

Gonad protection may be the key to species' survival in space. Sperm cells are the most sensitive cells to solar wind radiation, and on the Moon or in deep space, as there is no atmosphere, there is no protection like on Earth, so in a case of solar storm, there was fear of the possibility of remaining sterile or at least to easily undergo genetic mutations over time in their offspring. The other very sensitive organ is the thyroid, while the other organs are more resistant to damage from the electrons irradiated by the Sun, for example, the nails are absolutely not sensitive and therefore there is less need to protect them.

A study that measured blood indices during a spaceflight [2] on the second day reported that erythropoietin levels were significantly lower, indicating a suppression in erythropoiesis (Leach et al. 1988). Another study showed a reduction in erythropoietin levels within 24 h of exposure to microgravity (Udden et al. 1995). 11 Mar 2024

R. G. Russo, *Navigating Health Challenges in Space*,
https://doi.org/10.1007/978-3-032-03620-9_16

Erythropoiesis and Solar Eruptions

In addition to the hips, other sensitive areas are the shoulder blades, spine, femur, sternum, and skull. These bones in fact contain the erythropoietic bone marrow, i.e., "the blood factory" of the human body. The delicate cells of the bone marrow are particularly vulnerable to solar storms; a large dose of solar protons flowing through the body can destroy all erythropoietic cells. Without these cells that create new blood, or rather new cells of the red series, platelets and red blood cells (the latter also called erythrocytes), a person would become anemic within a week and could have an increased risk of bleeding.

The destruction of the bone marrow cells responsible for the production of monocytes, macrophages, and lymphocytes could increase the risk of contracting infections (particularly from fungi and bacteria that commonly live on the mucous membranes and skin). To survive the radiation of a solar storm, the highest priority must therefore be to protect the erythropoietic bone marrow. Surviving radiation emission during solar storms is more important than ever, especially in anticipation of future long-duration missions.

Outside the protection of the Earth's magnetic field and with no protective ambient atmosphere, an astronaut walking on the lunar surface, or working in space outside the orbiting station, is massively exposed to the mainstream of solar storms.

The best solution is to take shelter in appropriate structures, but if the shelter is too far away to be reached in time, it is necessary to bring a space suit with additional radiation protection that protects the areas rich in hematopoietic tissue such as the shoulders and spinal column; this can make the difference between living and dying.

Applying additional protection to the inside of the space suit may not be practical, however, because the space suit would become too bulky. Astronauts must be able to walk, jump, bend, and reach to grasp objects and tools. Too much protection would make these simple movements impossible, so it was decided to use selective protection: a layer of a plastic polymer such as polyethylene, only one centimeter thick, placed at key points, can prevent acute radiation sickness.

For all solar flares [3] (large eruptions of electromagnetic radiation from the Sun lasting from minutes to hours; the sudden outburst of electromagnetic energy travels at the speed of light, and therefore, any effect upon the sunlit side of Earth's exposed outer atmosphere occurs at the same time the event is observed), these measures are sufficient to keep the astronaut's hematopoietic system virtually intact. If only 5% of the bone marrow cells survive, the bone marrow will be able to regenerate and the person will survive, without the need for a transplant.

Solar flares usually take place in active regions, which are areas on the Sun marked by the presence of strong magnetic fields, typically associated with sunspot groups. As these magnetic fields evolve, they can reach a point of instability and release energy in a variety of forms. These include electromagnetic radiation, which are observed as solar flares.

An astronaut, thus shielded, could only develop long-term health problems, such as tumors, cataracts, and other late-onset diseases. No space suit can stop all the solar protons; however, if a sufficient supply of marrow remains intact, the astronaut can even survive long enough to deal with potential long-term damage. Reduced red cell production can result from the direct impairment of medullary erythropoiesis, as seen in the thalassemia syndromes, inherited bone marrow failure, and anemia in chronic diseases [4].

Alternatively, in disorders such as sickle cell disease (SCD) as well as enzymopathies and membrane defects, medullary erythropoiesis is not, or only minimally, directly impaired.

Despite these differences in pathophysiology, therapies have traditionally been nonspecific, limited to symptomatic control of anemia via packed red blood cell transfusion, resulting in iron overload and the eventual need for iron chelation or splenectomy to reduce defective red cell destruction.

Likewise, in Polycythemia Vera, Overproduction of Red Cells Has Historically Been Dealt with by Nonspecific Myelosuppression or Phlebotomy

Bibliography

1. Nandakumar SK, Ulirsch JC, Sankaran VG. Advances in understanding erythropoiesis: evolving perspectives. Br J Haematol. 2016;173(2):206–18.
2. Gilboa D, Haim-Ohana Y. Erythropoietin enhances Kupffer cell number and activity in the challenged liver. Sci Rep. 2017;7:10379.
3. Trudel G, Stratis D, et al. Transcriptomic evidence of erythropoietic adaptation from the International Space Station and from an Earth-based space analog. Microgravity. 2024;10:55.
4. De Santo NG, Cirillo M, et al. Anemia and erythropoietin in space flights. Semin Nephrol. 2005;25(6):379–87.

17 Alterations of the Immune System

What Does the Immune System Do?

The immune system works hard to keep you healthy. It does this by:

- Keeping invaders out of your body
- Destroying invaders
- Limiting how much harm the invaders can do if they are inside your body
- Healing damage to your body
- Adapting to new challenges and threats

Invaders of the immune system include:

- Bacteria
- Viruses
- Fungi
- Parasites
- Cancer cells

How Does the Immune System Work?

- Tells the difference between cells that are yours and those that do not belong in your body
- Activates and mobilizes to kill germs that may harm you
- Ends an attack once the threat is gone
- Learns about germs after you have had contact with them and develops antibodies against them
- Sends out antibodies to destroy germs that try to enter your body in the future

R. G. Russo, *Navigating Health Challenges in Space*,
https://doi.org/10.1007/978-3-032-03620-9_17

What Parts of the Body Make up the Immune System?

Many parts of the body, including immune system organs and cells, work together to keep you healthy. The main components of your immune system are:

- White blood cells.
- Antibodies.
- Cytokines.
- Complement system
- Lymph nodes
- Spleen
- Tonsils and adenoids
- Thymus
- Bone marrow
- Skin
- Mucosas

Conditions that Can Interfere with the Normal Workings of the Immune System

- Allergies
- Autoimmune diseases (e.g., lupus, rheumatoid arthritis)
- Primary immunodeficiency diseases
- Infectious diseases (e.g., HIV, mononucleosis)
- Cancer (e.g., leukemia, lymphoma)
- Sepsis

Signs and Symptoms of Immune System Disorders

They vary depending on the condition and may include:

- Always feeling tired (fatigue)
- Unexplained fever
- Unexplained weight loss
- Drenching night sweats
- Itchy skin
- Sore, aching muscles
- Fingers or toes that tingle or are numb
- Trouble concentrating or paying attention
- Hair loss
- Inflammation, rashes, or redness anywhere on the body
- Swollen lymph nodes in the neck, armpits, or groin

Radiation, microgravity, poor sleep, and other spaceflight stressors can alter the immune system of astronauts. Such space stressors may also trigger allergic reactions or awaken dormant viruses in astronauts' bodies, like those that cause cold sores and shingles. Studies on Earth previously indicated that shifts in immunity may lead to an increased incidence of infections and that elevated immunity may lead to allergies, so researchers hope to use Earth- and space-based research to better understand such irregularities as they prepare for longer-duration missions.

Immunosenescence is associated with complex alterations and dysregulation of immune function and inflammatory processes. Age-related changes in innate immune responses including alterations in chemotactic, phagocytic, and natural killing functions; impaired antigen presenting capacity; and dysregulated inflammatory response have been described [4].

Risk of Altered Immune System Responses

Radiation, microgravity, poor sleep, and other spaceflight stressors can alter the immune system of astronauts. Such space stressors may also trigger allergic reactions or awaken dormant viruses in astronauts' bodies, like those that cause cold sores and shingles. Studies on Earth previously indicated that shifts in immunity may lead to an increased incidence of infections and that elevated immunity may lead to allergies, so researchers hope to use Earth- and space-based research to better understand such irregularities as they prepare for longer-duration missions [5].

From the start of life on Earth, several immune defense mechanisms have evolved to guarantee cellular integrity, homeostasis, and host survival.

The Human Immune System in Space

Immune System's Basic Importance

Because space is not the place to have medical emergencies, experts are researching how spaceflight affects the immune system.

One strange effect is how the cells of the immune system operate in general. T cells and natural killer (NK) cells are the defenders of our immune system, having reduced function during spaceflights. That reduction can reactivate normally latent viruses. A dampened immune system also leads to mild illnesses, manifestation of allergies, and even rashes, which have been observed in a few astronaut cases. After testing astronauts' blood aboard the ISS, sampled before, during, and after their six-month sojourns, certain immune-related genes, about 100 of them, decline in activity when in space.

The innate immune system builds up the "first line of defense," consisting of elements such as mechanical barriers (skin and mucous layer), followed by neutrophils, macrophages, monocytes, acute phase proteins, cytokines, and the complement system on the cellular and molecular level 1. A fundamental characteristic of the adaptive immune system is the ability to distinguish what is "self" and what is "non-self." Missing foreign pathogens may lead to infection, and missing mutated own cells may lead to tumor formation; however, over-aggression of the immune system aimed at its own organism may lead to autoimmune diseases.

Astronauts in Future Deep-Space Missions

While work continues on aerospace rockets that enable manned missions to the Moon and Mars, we doctors ask ourselves what impact a prolonged stay in space could have on the human body. On August 25, 2023, an international research team provided the first answers to this question. Their exciting publication focused on the effects of space on the human immune system.

The research team also outlined a path to identify compounds that can reverse the effects of near-zero gravity and showed that one of these, quercetin, holds promise for mitigating damage caused by spaceflight and normal aging to the ground. "Our work provides a resource to better understand how and why the immune system changes in microgravity and simulated spaceflight conditions." "We also provide a way to develop countermeasures to maintain normal immunity under these challenging conditions."

Why Do Astronauts Get Sick More Easily?

Although fascinating and mysterious, space is, in fact, an extremely hostile environment for human health. One of the threats it poses is represented by the changes in the immune system that occur in astronauts during their stay in space, making them more vulnerable to infections and leading to the reactivation of latent viruses in the body.

Biological changes persist even after returning to Earth. For example, 50% of astronauts, returning to our planet, reported bacterial or viral infections, therefore suggesting some form of immune suppression in place at the time of landing.

Advances in space research confirm the profound impact of the space environment on adaptive immunity, including exposure to microgravity, increased radiation levels, psychological stress, and isolation in an artificial environment.

> If we want astronauts to be able to undertake safe space missions, we need to understand how their immune system is affected and try to find ways to counteract harmful effects on it.

The alteration of the immune system consists in a decrease in immune defenses. Initially, it was assumed that the anomaly was caused exclusively by radiation, but it was later verified that it is caused by various types of stress, such as confinement stress and suspension stress, generically called spaceflight stress.

Examples of autoimmune diseases are Addison's disease, celiac disease, dermatomyositis, Graves' disease, Hashimoto's thyroiditis, myasthenia gravis, multiple sclerosis, rheumatoid arthritis, Sjogren's syndrome, systemic lupus erythematosus, and type 1 diabetes.

Malfunction of the immune system may be triggered by external factors (for example, an infection, a non-tolerated drug), especially in people whose genes predispose them to autoimmune diseases. Autoimmune reactions can be triggered in different ways: alteration of a normal organic substance, for example, by viruses, drugs, sunlight, or radiation.

Astronauts in low Earth orbit, such as in the International Space Station, suffer conditions that attack the immune system, particularly infections, reactivation of latent viruses, and skin sensitivity. Long space journeys compromise the immune system of astronauts, mostly due to the lack of gravity. These reactions occur even during short spaceflights.

Consequences of Long-Term Exposure to Microgravity on the Immune System

The research team used blood samples from eight healthy volunteers to study the effects of microgravity on the immune system. Healthy volunteers were exposed to dry soaking for a total of 21 days. Dry immersion means immersing the subject in thermoneutral water, while covered by a special waterproof floating elastic fabric. The subject, wrapped in the fabric and freely suspended in the mass of water, remains dry. During horizontal immersion, the pressure forces are distributed almost equally over the entire surface of the body (only the head and neck are not entirely supported by the water). The absence of mechanical support of specific anatomical areas during immersion creates a state similar to weightlessness called "unsupportedness." Blood samples were collected 7 days before dry soaking; on days 7, 14, and 21 during dry soaking; and 7 days after dry soaking. The 21-day dry immersion was considered a long-term exposure to microgravity.

The research team managed to demonstrate the long-term effects of microgravity on the gene expression programs of T cells. T cells exposed to microgravity changed their transcription program quite rapidly. This is due to a downregulation of protein-coding and non-protein-coding genes responsible for controlling T cell activation and differentiation. At the same time, there was an upregulation of genes crucial for naïve T cells.

Decreased Immune System Functioning

Astronauts in space have weakened immune systems, which means that in addition to increased vulnerability to new exposures, viruses already present in the body—which would normally be suppressed—become active. In space, T cells do not reproduce properly, and the cells that do exist are less able to fight off infection. A NASA research is measuring the change in the immune systems of its astronauts as well as performing experiments with T cells in space.

In March 2019, NASA reported that latent viruses in humans may be activated during space missions, adding possibly more risk to astronauts.

Conclusion: The Immune System Weakens in Space

The absence of gravity led to an increase in the percentage of naïve T cells. At the same time, T cell differentiation was reduced. This could be associated with a longer activation phase of naïve T cells after contact with an antigen. This slows down the fight against infections. This could also explain the lower effectiveness against cancer cells. Based on the results of this study, it is now possible to develop methods to reverse these changes in the genetic program of immune cells.

Bibliography

1. Mahmood MM. A future life requires a super immune system. J Clin Res Rep. 2021;8(05) https://doi.org/10.31579/2690-1919/197.
2. Yu W, Yu Y, et al. Immune alterations with aging: mechanisms and intervention strategies. Nutrients. 2024;16(22):3830.
3. Marchal S, Choukér A, et al. Challenges for the human immune system after leaving Earth. Micrograviy. 2024;10:106.
4. Lewis ED, Wu D, Meydani SN. Age-associated alterations in immune function and inflammation. Prog Neuro-Psychopharmacol Biol Psychiatry. 2022;118:110576.
5. Smith LC, et al. In: Edwin LC, editor. Advances in comparative immunology. Springer International Publishing; 2018. p. 409–501.

Alterations of the Reproductive System 18

The female reproductive system has two functions: The first is to produce egg cells and the second is to protect and nourish the offspring until birth [2]. In addition to differences in nearly every reproductive organ, numerous differences typically occur in secondary sexual characteristics. The female reproductive system comprises internal and external organs that facilitate menstruation and procreation. This organ system is responsible for producing gametes (termed eggs or ova), regulating sex hormones, and maintaining fertilized eggs as they develop into mature fetuses ready for delivery. The male reproductive system has one function, and it is to produce and deposit sperm [3].

There is the possibility of infection and malfunction in the reproductive organs as well as the plausibility of sexual transmission of disease.

Alterations in Space

It has been estimated that exposure to space radiation (e.g., during a typical Mars mission) could decrease ovarian reserve by 50% by damaging primordial follicles. Furthermore, exposure to space radiation could also decrease the reproductive capacity of women astronauts by reducing the time interval for menopause onset.

Females

A previous study based on female mouse models has shown that microgravity adversely impacts the growth and differentiation of embryonic stem cells. This leads to the impairment of the decidualization of the endometrium, a vital factor for implantation and normal pregnancy.

Microgravity caused a 12% and 15% reduction in luteinizing hormone and progesterone, respectively. This could be due to dysfunction of *the corpus luteum.*

R. G. Russo, *Navigating Health Challenges in Space*,
https://doi.org/10.1007/978-3-032-03620-9_18

No change in menstrual cycle length was associated with microgravity. It must be noted that although abnormal uterine bleeding is a common problem of reproductive-aged women, research has not analyzed whether this problem is enhanced during space travel.

Typically, during spaceflight training or actual spaceflight, women astronauts undergo medically induced amenorrhea using combined oral contraceptives containing progestin and estrogen.

The aforementioned combination of oral contraceptives has been associated with higher concentrations of transferrin, lower circulating concentrations of albumin, and enhanced levels of inflammation markers. These conditions enhance the possibility of venous thromboembolism during spaceflight.

Disorders of the Female Reproductive System

Hormonal and Menstrual Alterations

- Amenorrhea (cessation of menstruation)
- Anovulation (failure of mid-cycle ovulation)
- Dysmenorrhea and premenstrual tension syndrome (painful menstruation)
- Dysfunctional uterine bleeding (uterine bleeding of undetermined cause)
- Polycystic ovarian syndrome

Alterations of Sexual Maturation

- Delayed puberty
- Precocious puberty

Sexual Dysfunction

- Disorders of sexual desire (decreased or increased)
- Anorgasmia (orgasmic dysfunction)
- Dyspareunia (painful intercourse)

Disorders of Fertility

Pelvic relaxation disorders

- Cystocele (bladder hernia), rectocele (rectum protrusion/drop), urethrocele (prolapse of the female urethra)
- Vaginal prolapse.
- Uterine prolapse.

Growths and proliferative conditions

- Endometriosis
- Benign ovarian cysts

- Endometrial polyps
- Leiomyomas
- Cancers: cervical, vaginal, endometrial, vulvar, and ovarian cancers

Males

In humans, therapeutic radiation higher than 1 Gy could lead to azoospermia. This level of radiation also enhances the risk of hereditary disorders.

When men with rectal cancer are treated with radiotherapy, their testes are often exposed to direct or scattered radiation, which in turn results in reduced serum testosterone levels.

Microgravity was found to impact the sperm motility negatively. However, during the transportation of male human gametes into space, it was observed that sperm integrity was protected by cryopreservation.

Disorders of the Male Reproductive System

- Disorders of the urethra
  - Urethritis
  - Urethral strictures
- Disorders of the scrotum and testes
  - Varicocele (varicose veins in the scrotum; rarely painful but can lead to infertility)
  - Hydrocele (fluid in the scrotum; a painless swelling of one or both testicles)
  - Spermatocele (painless fluid filled cyst in the epididymis)
  - Testicular torsion (twisting of the spermatic cord → severe testicular pain)
  - Epididymitis (inflammation of epididymis → severe testicular pain)
- **Disorders of Fertility**
- **Disorders of the Prostate**

- Benign prostatic hyperplasia.
- Prostate cancer
- Prostatitis

- **Erectile Dysfunction**
- **Disorders of the Breast**
  - Gynecomastia
  - Breast cancer

Alterations of the Reproductive System

Travel in deep space poses multiple simultaneous hazards to human reproduction that are beginning to be understood. Recently, scientists have reviewed the existing literature to understand how space travel, space radiation, and microgravity affect human reproductive functions.

The rigors of spaceflight including radiation, gravitational stressors, and circadian rhythm changes could potentially affect ovulation, fertilization, endometrial receptivity, preimplantation embryo development, embryo implantation, placentation, and pregnancy.

In vitro it has been seen that the spermatozoon and the egg, when they begin to multiply, are unable to form an orderly subdivision structure, so the distribution of cellular matter and chromosomes occurs in a chaotic manner, conceptually comparable to tumor production.

In addition, it affects bone and muscle mass, alters electrolyte balance and the cardiovascular system, and enhances glomerular filtration rate in both sexes. Chronic immunosuppression remains a necessity to prevent organ rejection, despite increased risks of infection, organ toxicity, and malignancies. Abnormalities of female gonadal function in patients of reproductive age are recognized; however, pathological alterations of the reproductive system in patients treated with new-generation immunosuppressive drugs are still poorly documented.

Our findings suggest that pre- and post-transplant evaluation of female patients should include menstrual history, baseline pelvic ultrasound, and hormonal levels to assess the presence and monitor the progression of such alterations.

Because we know women can have their periods in space, it is definitely possible for ovulation to occur. This means that the lining of the womb has grown, so it is preparing for pregnancy. Therefore, technically fertilization might occur as well. Whether that is actually possible is a real unknown.

In the future, space radiation's effect on fresh and frozen semen samples must be studied. This study could provide insights into the possibility of developing a human sperm bank outside the Earth.

Future Outlook

In the case of female astronauts, pilots, cabin crews, and frequent flyers, the radiation may exceed the recommended level. Hence, it is important to understand the impact of excess radiation exposure on their overall health.

Furthermore, studying the impact of multi-year duration spaceflights or colonization on pregnancy, particularly fetal growth, is imperative.

The fate of the offspring conceived and developed in space must be analyzed in future research.

Understanding whether microgravity and space radiation impacts the pro-oxidant/antioxidant balance during pregnancy is important. An altered balance could increase the risk of miscarriage, preterm birth, and inappropriate fetal growth.

In the future, space radiation's effect on fresh and frozen semen samples must be studied. This study could provide insights into the possibility of developing a human sperm bank outside the Earth..

Bibliography

1. Gomes C, da Rocha R, da Rocha C, et al. Alterations of the male and female reproductive systems induced by COVID-19. Wien Klin Wochenschr. 2021;133(17–18):966–72.
2. Rosner J, Samardzic T, Sarao MS. Physiology, female reproduction. StatPearls [Internet]; 2024.
3. Swaab DF, Garcia-Falgueras A. Sexual differentiation of the human brain in relation to gender identity and sexual orientation. Funct Neurol. 2009;24(1):17–28.

19 Alterations of the Nervous System

The nervous system [1] can be affected by various disorders. It can be damaged by:

- Injury
- Infections
- Degeneration
- Structural defects
- Tumors
- Blood flow disruption
- Autoimmune disorders

Disorders of the Nervous System [2]

- Vascular disorders, such as stroke, transient ischemic attack (TIA), subarachnoid hemorrhage, subdural hemorrhage and hematoma, and extradural hemorrhage
- Infections, such as meningitis, encephalitis, myelitis, and epidural abscess
- Structural disorders, such as brain or spinal cord injury, cervical spondylosis, carpal tunnel syndrome, and brain or spinal cord tumors
- Seizure disorders, such as epilepsy
- Degeneration, such as Parkinson's disease, amyotrophic lateral sclerosis (ALS), Huntington's chorea, and Alzheimer's disease
- Autoimmune or inflammatory disorders, such as Bell's palsy, multiple sclerosis, peripheral neuropathy, and Guillain-Barré syndrome
- Mental health disorders, such as mood disorders, depression, and schizophrenia

Moreover, the altered nervous system symptoms and consequences may be due to:

R. G. Russo, *Navigating Health Challenges in Space*,
https://doi.org/10.1007/978-3-032-03620-9_19

- *Guillain-Barré syndrome*, a disorder in which the immune system attacks the body itself, specifically the nervous system. Its most characteristic symptoms include a tingling sensation in the legs and muscle weakness.
- *Neurological diseases* such as Parkinson's, whose early symptoms are often tremors and twitching in the limbs; or multiple sclerosis, in which tingling, numbness, or weakness may appear, among many other symptoms.
- When the problem is caused by *nerve compression*, depending on where the nerve is compressed, disorders such as carpal tunnel syndrome or thoracic outlet syndrome may occur.
- Nervous system disturbance can also be due to, for example, *alcoholism*.
- Other conditions that can disrupt the nervous system include *cerebral vascular disorders, epilepsy, meningitis, concussions, and traumatic brain injury*, among others.

These are the *most common symptoms* of a nervous system disorder. But each person may have slightly different symptoms. Symptoms may include:

- Persistent or sudden onset of a headache
- A headache that changes or is different
- Loss of feeling or tingling
- Weakness or loss of muscle strength
- Loss of sight or double vision
- Memory loss
- Impaired mental ability
- Lack of coordination
- Muscle rigidity
- Tremors and seizures
- Back pain that spreads to the feet, toes, or other parts of the body
- Muscle wasting and slurred speech
- New language impairment (expression or comprehension)

Alterations of the nervous system are not as significant as muscle wasting and osteoporosis, but they cause problems that have repercussions both on work and in relationships with colleagues.

In fact, they might include space motion sickness, vestibular dysfunction, and spaceflight-associated neuro-ocular syndrome. Spaceflight also has the potential to cause longer time complications such as an increased risk of brain malignancies, premature aging, and accelerated neurodegeneration.

Moreover, several neurologic alterations, including ataxia, postural disturbances, perceptual illusions, neuromuscular weakness, fatigue, and ocular changes have been attributed to the effects of microgravity.

The neuro consequences of the spaceflight environment result in headward fluid shifts, ventricular expansions, an upward shift of the brain within the skull, and remodeling of gray and white matter. The fluid changes are correlated with changes to perivascular space and spaceflight-associated neuro-ocular syndrome.

Numerous studies have reported macrostructural changes to the brain after spaceflight, including alterations in brain position, tissue volumes, and cerebrospinal fluid distribution and dynamics. Changes in brain tissue microstructure and connectivity were also described, involving regions related to vestibular, cerebellar, visual, motor, somatosensory, and cognitive functions. Several alterations were also associated with exposure to analogs of spaceflight, providing evidence that brain changes likely result from cumulative exposure to multiple independent environmental stressors.

Bibliography

1. Thau L, Reddy V, et al. Anatomy, central nervous system. StatPearls [Internet]; 2022.
2. Gupta U, Baig S, et al. The neurology of space flight; how does space flight effect the human nervous system? Life Sci Space Res. 2023;36:105–15.

Alterations of Kidney Function 20

The body takes nutrients from food and converts them to energy. After the body has taken the food components that it needs, waste products are left behind in the bowel and in the blood. The kidney and urinary systems help the body eliminate liquid waste called urea and keep chemicals, such as potassium and sodium, and water in balance. Urea is produced when foods containing protein, such as meat, poultry, and certain vegetables, are broken down in the body. Urea is carried in the bloodstream to the kidneys, where it is removed along with water and other wastes in the form of urine [1].

Spaceflight induces (1) renal transporter dephosphorylation which may indicate that astronauts' increased risk of nephrolithiasis is in part a primary renal phenomenon rather than solely a secondary consequence of bone loss; (2) remodeling of the nephron that results in expansion of distal convoluted tubule size but loss of overall tubule density; and (3) renal damage and dysfunction when exposed to a Mars roundtrip dose-equivalent of simulated GCR (galactic cosmic radiation) [2].

Alterations of the Urinary System in Space

Knowledge of the renal response during space missions remains limited. The impact on kidney function is multifactorial, and besides, microgravity and radiation exposure includes altered caloric and hydration intake compared to terrestrial circumstances, as well as multiple neurohormonal changes.

Renal alterations, encompassing a wide spectrum of renal and urinary disorders, are a common yet complex area of focus for nursing professionals. Whether it is nephritis, nephrolithiasis, or renal failure, these conditions can profoundly impact the health and quality of life of clients. Understanding the pathophysiology of these disorders, incorporating evidence-based practice, and implementing a client-centered approach are crucial for delivering effective nursing care [3].

R. G. Russo, *Navigating Health Challenges in Space*,
https://doi.org/10.1007/978-3-032-03620-9_20

In microgravity, the cephalad redistribution of blood volume with the concomitant decrease in the pressure gradient should reduce kidney perfusion. In turn, a lower perfusion pressure within the renal artery should lead to a greater release of renin and activation of the renin/angiotensin/aldosterone system.

Moreover, microgravity may have a direct effect on the crystallization and nucleation of nascent kidney stones.

Urine collection devices build up the bacterium *Proteus mirabilis*, which is associated with urinary tract infection. For this reason, astronauts may be susceptible to urinary tract infection [4].

Urinary System in Space

Exposure to the altered gravity in the spaceflight environment may cause physiological changes. One of these changes is the inability to completely empty the bladder or urinary retention.

Causes of urinary retention in the early phases of flight include altered baseline physiology seen with exposure to microgravity, the anticholinergic side effects of medications that are taken to combat space motion sickness, and other factors. Urinary retention may impact health on orbit by causing discomfort and increasing the risk of urinary tract infection.

Loss of Kidney Functiom [5]

"Serious health risks emerge (with respect to the kidneys) the longer a person is exposed to Galactic Radiation and microgravity." In fact, based on their current research with mice, the researchers predicted that astronauts who have been exposed to microgravity, reduced gravity, and galactic radiation for 3 years or so on a Mars mission may have to return to Earth while attached to dialysis machines.

Bibliography

1. Bibliography. Current world literature. Genitourinary system. Curr Opin Oncol. 1999;
2. Barrell EA, Burton AJ. Alterations in urinary function. In: Large animal internal medicine; 2020. p. 166–173.e1.
3. Volpin F, Badeti U, Wang C, et al. Urine treatment on the international space station: current practice and novel approaches. Membranes (Basel). 2020;10(11):327.
4. Law J, Cole R et al. (Johnson Space Center) Aerospace Medical, 2015.
5. Vaidya SR, Aeddula NR. Chronic kidney disease. StatPearls [Internet] July 31; 2024.

Dental Material and Oral and Stomatological Diseases

21

Dental materials include such items as resin composites, cements, glass ionomers, ceramics, noble and base metals, amalgam alloys, gypsum materials, casting investments, dental waxes, impression materials, denture base resins, and other materials used in restorative procedures.

Currently, little is known and much more research needs to be done with an emphasis on the interdisciplinary aspect. In order to achieve good oral health, we must seek collaboration between nutritionists, engineers, flight surgeons, and most of all dental specialists, all experts in their own field [1]. The dental hygiene module aims to impart to the student the knowledge regarding the tools and techniques to carry out and maintain proper oral hygiene at home, as well as to acquire the skills to transmit these notions to the patient, motivating her/him in order to maintain over time a good level of oral hygiene.

The Oral Diagnosis Principles module provides knowledge about the main pathological conditions of the oral cavity, with reference to the underlying pathophysiological mechanisms.

Exploring the complex nature of performing dental procedures in microgravity, where issues such as aspiration and blood clotting pose significant challenges, the focus on preventive measures takes on paramount importance in maintaining optimal oral health for astronauts enduring missions lasting several years. The different unique constraints of communication with Earth, characterized by significant delays due to vast distances, also play a significant role [2].

The dental guidelines for astronauts are beginning to take shape as we explore all the different aspects of maintaining oral and dental health in space.

R. G. Russo, *Navigating Health Challenges in Space*,
https://doi.org/10.1007/978-3-032-03620-9_21

Bibliography

1. Zhu J, Wenlin C, Jun L, et al. Dental materials for oral microbiota dysbiosis: an update. Front Cell Infect Microbiol. 2022;
2. Samathu JHA, Mani R. Smile beyond the stars: a narrative review exploring the challenges for dentistry in space. Cureus. 2014;16(8)

22 Impairment of Eyesight

During long spaceflight missions, astronauts may develop ocular changes and visual impairment collectively known as the space-associated neuro-ocular syndrome (SANS). Such vision problems may be a major concern for future deep space flight missions, including a human mission to Mars [1].

Microgravity causes a person's blood and cerebrospinal fluid to shift toward the head, and researchers suspect that these fluid shifts are an underlying cause of SANS.

Crew members aboard the International Space Station conducted a variety of scientific investigations, including conducting sessions for **ISAFE**, which measures changes in the astronauts' eyes, brain, and blood vessels during missions of varying lengths. ISAFE measurements could help scientists determine whether SANS varies with mission length, whether vision recovers after returning to Earth, and how long recovery takes. That insight is needed to protect the eyes and vision of astronauts on future long-duration spaceflights. SANS shares some characteristics with conditions on Earth, including increased pressure in the brain, and results also could provide new insights benefiting people who experience those conditions [2].

In another experiment, Ocular Health, researchers found structural alterations to the eyes that occurred early during spaceflight and persisted throughout the mission. Postflight recovery took 30–90 days. Some of the analytic tools used for ISAFE are enhancements of those used for Ocular Health.

Spending long periods in the microgravity of space can lead to changes in the eye, including swelling in the region where the optic nerve extends to the brain, flattening of the rear of the normally round organ, emergence of wrinkles at the back of the retina, and shifts in the refractive index that change how the eye functions.

R. G. Russo, *Navigating Health Challenges in Space*,
https://doi.org/10.1007/978-3-032-03620-9_22

Several lines of evidence suggest that a simulated or actual space environment can induce changes in circadian rhythms [36, 37, 41–44]. Independent orbital studies have demonstrated that the amplitude of body temperature rhythmicity decreases during spaceflight in comparison with that on the ground.

Bibliography

1. Science in Space: July 28, 2023 – Astronaut Vision, JUL 31, 2023.
2. Savage N. Mysterious syndrome remains a 'red risk' for long-term spaceflight. 05 March 2025.

23 Alteration in the Daily Circadian Rhythm

The shuttle orbits the Earth at such a high speed, approximately 27,700 km/h at 354 kilometers above sea level, that the Sun can be seen rising and setting at least 15 times in the course of 24 h. This phenomenon, which can be fascinating, quickly turns into "torture," because the Sun can rise when you are going to bed or while you are sleeping. The Sun can rise at least two or three more times in the space of a few hours. Sleep disturbances lead to serious problems, such as inattention and mood disorders [1].

The human body has a natural automatic clock called the circadian rhythm. When it does not work right, it can throw you out of sync with sleeping, waking up, and checking the time of day. Most of these disorders are treatable, and some are temporary and resolve with simple changes and adjustments. These disruptions affect how well you sleep, when you sleep, and how you function while awake. They are also known as "circadian rhythm sleep disorders" or "circadian rhythm sleep-wake disorders." Experts classify them as "circadian sleep-wake disorder not otherwise specified."

The symptoms of these various disorders may be:

- Trouble falling asleep (insomnia)
- Difficult staying asleep
- Trouble falling back asleep during the middle of the night or early in the morning
- Waking up earlier than you want or plan
- Trouble waking up (sleep inertia) or unpredictable sleep-wake times
- Sleepiness during the day, which can lead to hypersomnia
- Needing to nap during the daytime
- Fatigue
- Headaches
- Depression, irritability, and other issues due to lack of sleep

R. G. Russo, *Navigating Health Challenges in Space*, https://doi.org/10.1007/978-3-032-03620-9_23

Furthermore, it should be considered that the human skin, rich in melatonin, is sensitive to light, and it is this that regulates the circadian rhythm. At dawn, light gives the awakening signal; therefore, sleep can be very disturbed when the light-dark circadian rhythm is not eactly 24 h. The other problem linked to sleep is that the body is used to falling asleep when it is lying down and the muscles relax, while in microgravity, this sensation is absent. Furthermore, astronauts are forced to tie themselves to the bed with straps (a factor which may be unbearable for some astronauts) to be able to sleep without the risk of moving while floating in the air [2].

Bibliography

1. Guo J-H, Qu W-M, Chen S-G, et al. Keeping the right time in space: importance of circadian clock and sleep for physiology and performance of astronauts. Mil Med Res. 2014;
2. Gundel A, Polyakov VV, Zulley J. The alteration of human sleep and circadian rhythms during spaceflight. J Sleep Res. 1997;1:1–8.

Alterations in Relationship Life

24

Research on the Effects of Staying in Space

For a study on the effects of staying in space on the human organism (especially radiation at the genetic level), a US astronaut and a Russian cosmonaut spent a year on the International Space Station, from March 27, 2015, to March 2, 2016.

Alterations in Relationship Life

Relationships naturally change as they grow and develop. Dealing with changes starts with understanding your partner and learning ways to manage differences.

Astronauts exhibit several pathophysiological changes due to a variety of stressors related to the space environment, including microgravity, space radiation, isolation, and confinement. Space motion sickness, bone and muscle mass loss, cardiovascular deconditioning and neuro-ocular syndrome are some of the spaceflight-induced effects on human health. Optimal nutrition is of the utmost importance and—in combination with other measures, such as physical activity and pharmacological treatment—has a key role in mitigating many of the above conditions.

Bibliography

1. Chaloulakou S, Poulia KA, Karayiannis D. Physiological alterations in relation to space flight: the role of nutrition. Nutrients. 2022;14(22)

R. G. Russo, *Navigating Health Challenges in Space*,
https://doi.org/10.1007/978-3-032-03620-9_24

Part V

Other Related Information

Doctors in Space

25

What Do We Have to Study to Go into Space?

The first thing you need to know to become an astronaut is that there is no real course of study or school for astronauts. Astronaut candidates represent a wide spectrum of professionals such as test pilots, engineers, scientists, educators, and physicians. Candidates who have considerable experience working in a team setting and who habitually work outside their comfort zones are highly regarded. Like prospective medical students, astronaut candidates complete questionnaires and submit university transcripts, letters of reference, and a personal essay ("why I want to be an astronaut") as part of the application process.

Following selection (a gratifying day for a fortunate few!), astronauts begin the first phase of our education called basic training. This phase typically lasts a couple of years [1].

Astronaut requirements [2] have changed with NASA's goals and missions. Today, to be considered for an astronaut position, applicants must meet the following qualifications:

1. Be a US citizen
2. Possess a master's degree* in a STEM [3] field, including engineering, biological science, physical science, computer science, or mathematics, from an accredited institution
3. Have at least 2 years of related professional experience obtained after degree completion or at least 1000 h pilot-in-command time on a jet aircraft
4. Be able to pass the NASA long-duration flight astronaut physical

*The master's degree requirement can also be met by:

- Two years (36 semester hours or 54 quarter hours) of work toward a doctoral program in a related science, technology, engineering, or math field

R. G. Russo, *Navigating Health Challenges in Space*,
https://doi.org/10.1007/978-3-032-03620-9_25

- A completed Doctor of Medicine or Doctor of Osteopathic Medicine degree
- Completion (or current enrollment that will result in completion by June 2021) of a nationally recognized test pilot school program

Astronaut candidates must also have skills in leadership, teamwork, and communications.

However, to be considered for an astronaut position today, applicants must have the following qualifications:

- U.S. citizenship;

Master's degree in a STEM [4] field (including engineering, biology, medicine, physics, chemistry, geology, mathematics, or computer science from an accredited institution). Science, technology, engineering, and mathematics (STEM) is a term used to group together a few of the technical disciplines of science, technology, engineering, and mathematics. The term is typically used in the context of education policy of curriculum choices in schools. It has implications for workforce development, national security concerns (as a shortage of STEM-educated citizens can reduce effectiveness in this area), and immigration policy, with regard to admitting foreign students and tech workers. There is no universal agreement on which disciplines are included in STEM.

The Master's Degree Requirement Can Also Be Met By

Two years (36 semester hours or 54 quarter hours) of work toward a doctoral program in a related science, technology, engineering, or math field. A completed Doctor of Medicine or Doctor of Osteopathic Medicine degree. Completion (or current enrolment that will result in completion by June 2021) of a nationally recognized test pilot school program. Astronaut candidates must also have skills in leadership, teamwork, and communications.

Space Exploration for the Benefit of Medicine

Aviation and space medicine offers an extensive array of career opportunities for medical professionals, encompassing wide areas of practice in either a civilian or military capacity.

The outer space sector serves as a catalyst for advancing medical knowledge and technology transfer, yielding new vaccines, healthcare tools, and procedures. Satellite communications enhance telemedicine and telehealth, connecting medical experts with underserved regions.

When space innovation meets the fields of medicine and life sciences, the result is spectacular: advanced technologies developed in space exploration, especially by the US space agency NASA, originally designed for the agency's own astronauts and then commercially applied on land, have been revolutionizing the medical sector for decades.

Benefits of Space for Global Health

1. *Telemedicine and telehealth*: Space technology, including satellites, enables telemedicine and remote healthcare services. Remote areas and disaster-stricken regions can benefit from medical consultations, diagnoses, and treatment advice delivered by healthcare professionals from afar. Satellite-based communication systems facilitate real-time interactions between doctors and patients, improving access to healthcare services (see Chap. 33).
2. *Tele-epidemiology for epidemic and disease monitoring*: Satellites and space-based sensors can be used to monitor the spread of diseases, track vector populations (such as mosquitoes carrying diseases like malaria), and assess environmental factors that contribute to disease outbreaks. This data can aid in early detection and response to epidemics.
3. *Spin-off benefits of space research and development*: Medical imaging and diagnostics—Space technology has led to advancements in medical imaging and diagnostics. Technologies developed for space missions, such as high-resolution imaging devices, have been adapted for medical use. Imaging techniques like magnetic resonance imaging (MRI) and computed tomography (CT) owe some of their advancements to innovations in space technology. Health monitoring wearables—The miniaturization of sensors and electronics used in space missions has paved the way for wearable health monitoring devices. These devices can track vital signs, activity levels, and other health metrics, helping individuals manage their health and enabling healthcare providers to monitor patients remotely. Vaccine development and drug testing—Microgravity environments in space can provide insights into molecular structure and biological processes that are difficult to replicate on Earth. This has implications for vaccine development and drug testing. Space experiments have helped researchers understand how microorganisms behave and how human cells respond, leading to more effective drug development.
4. *Disaster management and humanitarian aid*: Space technology assists in disaster management and humanitarian aid efforts by providing rapid and accurate information about disaster-affected areas. This information helps coordinate relief efforts, assess infrastructure damage, and plan medical response operations effectively. Water and sanitation management—Remote sensing technology, often used in space applications, can help monitor water quality and identify potential sources of contamination. This information is crucial for ensuring access to clean and safe drinking water, thereby preventing waterborne diseases and improving global public health.
5. *Human development through health education and outreach*: Space agencies often engage in public outreach and education programs to promote interest in science, technology, engineering, and mathematics (STEM) fields. These pro-

grams can inspire students to pursue careers in healthcare, leading to an increased pool of skilled professionals.

Impact on Science and Medicine

Astronauts are not the only ones who benefit from space medicine research. Several medical products have been developed that are space spin-offs, which are practical applications for the field of medicine arising from the space program. Because of joint research efforts between NASA, the National Institutes on Aging, and other aging-related organizations, space exploration has benefited a particular segment of society—the seniors. These spin-offs are sometimes termed as "exomedicine."

Quite a few very useful inventions have been created in the field of medicine and medical instrumentation with NASA's help. Below are a few examples:

- *Personal alert systems*: These are emergency alert devices that can be worn by individuals who may require emergency medical or safety assistance.
- *CAT and MRI scans*: Their development would not have been possible without the technology recently provided by NASA.
- *Orthopedic evaluation tools*: Equipment to evaluate posture, gait, and balance disturbances was developed at NASA, along with a radiation-free way to measure bone flexibility using vibration.
- *Foam cushioning*: Special foam used for cushioning astronauts during liftoff is used in pillows and mattresses at many nursing homes and hospitals to help prevent ulcers, relieve pressure, and provide a better night's sleep.
- *Kidney dialysis machines*: Together with NASA, it has been developed, in the late 1960s, a system that would purify and recycle water during space missions.
- *Talking wheelchairs:* These were built to help paralyzed individuals who have difficulty speaking. Developed by NASA to create synthesized speech for aircraft, this technology is portable and versatile, as well as a highly successful speech prosthesis.
- *Surgically implantable heart pacemakers:* These devices depend on technologies developed by NASA for use with satellites.
- *Implantable heart defibrillator:* This tool continuously monitors heart activity and can deliver an electric shock to restore heartbeat regularity.
- *EMS communications:* This technology used to communicate telemetry between Earth and space was developed by NASA to monitor the health of astronauts in space.
- *Weightlessness therapy:* It can help some individuals with limited mobility on Earth and normally confined to wheelchairs. This idea also led to the development of the anti-gravity treadmill from NASA technology.
- *Diabetic foot mapping:* This technique was developed at NASA's center in Cleveland, Ohio, to help monitor the effects of diabetes in feet.

- *Collapsible, lightweight wheelchairs:* These were designed for portability and can be folded and put into trunks of cars. They rely on synthetic materials that NASA developed for its air and spacecraft.
- *Implantable heart defibrillator:* This tool continuously monitors heart activity and can deliver an electric shock to restore heartbeat regularity.
- *Weightlessness therapy:* The weightlessness of space can allow some individuals with limited mobility on Earth—even those normally confined to wheelchairs—the freedom to move about with ease.
- *Anti-gravity treadmill from NASA technology:* This employs "differential air pressure to mimic gravity."

We Are Made to Travel in Space: This Is How Our Body Reacts and Adapts

Astronauts in space play a crucial role in scientific research on human health. During space missions, experiments and research conducted in orbit provide important information on cardiovascular, metabolic, and musculoskeletal conditions. For example, women tend to lose consciousness earlier than men upon reentry, and astronauts can suffer from eye problems. The absence of gravity impacts the bones, leading to an osteoporotic process. Therefore, astronauts follow a specific diet with vitamin D and calcium.

Furthermore, spaceflight induces changes in the organism similar to those of a longer stay, such as elevated levels of cytokines and changes in gene expression linked to the activation of the immune system and DNA repair. These challenges represent a fertile ground for medical research and will require significant advances in medicine to address long-duration space journeys to the Moon or Mars. The spaces themselves, through various elements, contribute to the health of patients by reducing stress and improving the working conditions of healthcare personnel.

The general effects of spaceflight or reduced gravity on the body, for example, include changes in physical factors such as body weight, fluid pressure, *convection, and sedimentation.* [**]These changes affect body fluids, gravity receptors, and weight-bearing structures. The body will adapt to these changes over time spent in space. Travel in the confined space of a spacecraft can also induce psychosocial changes.

Convection is the process of transforming heat through air or liquid contents. Convection causes liquid or gas to heat up, expand, and decrease in density. This causes movement of the liquid or gas in a convection current. Furthermore, energy is passed from one particle to another upon contact, transferring heat.

Sedimentation is the process of separating small particles and sediments in water. This process happens naturally when water is still because gravity will pull the heavier sediments down to form a sludge layer. However, this action can be artificially stimulated in the water treatment process. Sedimentation is one of the processes through which we can remove high impurities present in a solution.

During Space Exploration [5], the use of astronomy and space technology to explore outer space is highly valuable, but it is always recommended to maintain maximum attention when using machinery. In fact, astronauts risk contact dermatitis, particularly if they are exposed to skin sensitizers such as acrylates. This could jeopardize a mission unless appropriate measures are taken to identify the source of the exposure, assess the health risks, and mitigate the exposure.

Air Pollution is the single most important cause of environmental health hazards. It can cause lung and heart diseases and some types of cancer, all resulting in premature deaths [6]. Toxic, or hazardous, air pollutants are substances that cause or are suspected of causing cancer, birth defects, or other serious harms. They can be gases, such as hydrogen chloride, benzene, and toluene or compounds and metals such as asbestos, cadmium, mercury, and chromium. The US Environmental Protection Agency has classified 188 pollutants as hazardous.

Air Contaminants such as air pollution that results from manufacturing and production have been exacerbated by human and technological systems. Air pollution results in negative health consequences such as damage to the respiratory, neurological, immune, and reproductive systems of humans who live in heavily polluted areas.

Toxic Waste is created by the manufacturing and production processes. It contains harmful chemicals that impact the health of plants and animals that come into contact with the waste in their environment. Although toxic waste is typically disposed of using regulated disposal methods, it impacts humans when waste seeps into the ground and enters the water table, contaminating the water source.

Air and Water Pollution are primarily a result of technological advances in transportation, manufacturing processes, and production methods.

Chemical Hazards can be gases, such as hydrogen chloride, benzene, and toluene or compounds and metals such as asbestos, cadmium, mercury, and chromium. The most common chemical hazards found in the workplace are asphyxiants, corrosives, irritants, sensitizers, carcinogens, mutagens, and teratogens.

Heavy Metals are naturally occurring elements but are used in trace amounts in man-made products such as vaccinations. Overexposure to heavy metals has been linked to health disorders such as neurological degeneration and increased cancer risks.

Pesticides are human-made poisons. As more equipment is required, environmental hazards in space may arise due to food production. Pesticides are important because they increase food production abilities, but they pose an environmental threat because they contain chemicals that are designed to kill insects and other pests that consume raw food. These chemicals are harmful to humans when consumed in large doses.

Radiation Hazards [7] Space radiation consists of high-energy particles such as protons, alpha particles, and heavier particles originating from sources such as galactic cosmic rays, energetic solar particles from solar flares, and trapped radiation belts. Space station crew exposures will be much higher than those on Earth, and unshielded astronauts may experience serious health effects if unprotected. Galactic cosmic radiation is extremely penetrating, and it may not be possible to build shields of sufficient depth to prevent or control exposure [8].

Radiation hazards may also come from man-made sources, for example, medical investigations, radio-isotopic power generators, or small experiments as on Earth. Astronautical hygienists will need to assess the risks from these other sources of radiation and take appropriate action to mitigate exposure.

Radiation Exposure is an environmental hazard that results in increased cancer risks. An acute exposure event can lead to radiation sickness which includes symptoms such as nausea, vomiting, and death if the acute exposure is prolonged. Repeated exposure to low levels of radiation increases the risk of developing cancer.

Trapped Radiation, Galactic Cosmic Radiation, and Solar Particle Events Are Biological Hazards [9]

In space, astronauts may face decreases in blood volume, orthostatic tolerance, and aerobic capacity while also experiencing increased arrhythmias. Although the cardiovascular system functions well in space, the body does not require as much work from the heart (still a muscle, after all) in microgravity.

Unseen Hazards [10] These are objects in space that emit types of electromagnetic radiation to which the human eye is insensitive. These radiations include radio waves, infrared and ultraviolet radiation, and x rays and gamma rays.

Ergonomic Hazards [11] Ergonomics is the science of planning and organizing products, systems, and environments to make them more comfortable, efficient, and safe for human use. In interior design, ergonomics involves creating spaces that promote good posture, reduce physical strain and stress, and promote overall well-being. These include repetition, awkward posture, forceful motion, stationary position, direct pressure, vibration, extreme temperature, noise, and work stress.

Electrical Hazards [12]The five main electrical hazards are electric shock and burns from contact with live parts, injury from exposure to arcing (when electricity jumps from one circuit to another), fire from faulty electrical equipment or installations, explosion caused by unsuitable electrical apparatus, and static electricity igniting flammable vapors or dusts, for example, in a spray-paint booth.

Noise Noise exposure in spaceflight can negatively impact hearing, but it can also cause other physiological effects (e.g., irritation, headaches, and degradation in sleep and relaxation). Fans, compressors, motors, transformers, pumps, etc., on the International Space Station (ISS) all generate noise.

In addition to health risks, loud noises may cause danger warning alarms to be heard.

Effects of Health in Space

From a medical perspective, long-duration spaceflight has physiological impacts on astronauts. Accelerated bone decalcification, similar to osteopenia and osteoporosis on Earth, is just one such condition. Another serious concern is the effects of space travel on the kidneys. Current estimates of the renal effects indicate that unless some additional remedial strategy against kidney damage is employed, astronauts exposed to microgravity, reduced gravity, and galactic radiation for 3 years or so on a Mars mission may have to return to Earth attached to dialysis machines.

Mankind in Space

There is no specific border between the Earth's atmosphere and space, but in most cases, space is beyond 100 kilometers where air rarely exists. If you stay for a long time in space, your muscles and bones will weaken, primarily in the legs and lower back. Gravity always acts on you while you are on Earth, so even if you are not really conscious of resisting gravity, you are always using the muscles of your lower body. In space, where gravity is very weak, posture can be maintained without standing on your legs, and there is no need to use your legs to move about. Therefore, research is underway to verify, in space, the effects of existing drugs for preventing bone loss by astronauts. Also, to prevent weakening of muscle and bone, astronauts exercise for about 2 h a day during their stay on the International Space Station (ISS).

What Would Happen to the Human Body in the Vacuum of Space?

What would happen to your body if it were exposed to the vacuum of space if you were not wearing a space suit? Astronauts floating without a suit in space would not survive, but their demise would happen within minutes, not within seconds, and it would be a gnarly exit, with boiling bodily fluids and a nearly frozen nose and mouth.

With more than 20 years of operations in low Earth orbit, NASA is taking a stepping stone approach to human exploration in space. Considering the extensive training and screening astronauts go through to ensure they are in optimal health for missions, it may seem ironic that they are also more likely to get sick while in space.

Different kinds of clinical events manifest in space, according to NASA. There are colds, sore throats, abnormal allergies, and even skin rashes. On the ISS, or spacecraft, the force of gravity is much weaker than on the earth. As a result, many astronauts show symptoms of "space sickness," such as headaches, nausea, and vomiting—especially nausea-inducing motion sickness attributed to a condition called space adaptation syndrome.

Conclusion The surface of the earth is enveloped by the atmosphere. This atmosphere provides the oxygen we need to breathe and also protects organisms from the UV rays and radiation that impinge on Earth. Astronauts who stay in space, where there is almost no atmosphere, are exposed to higher energy radiation than on Earth. If a person is exposed to a lot of higher energy radiation, the risk that they will develop diseases such as cancer increases.

Physical Health Is Not the Only Thing at Risk

To state the obvious, when our bodies feel bad, we feel bad, too. "The other thing to think about in terms of illness is how it affects one's mental performance and psychological and emotional state."

The ISS is about 250 miles above sea level, but there are no quick ways home from space. A study by NASA in 2010 analyzed ten astronauts' journals, allowing a deep dive into their psyche, morale, and behavior over long-duration spaceflight. Crew members wrote about things spanning from their workloads, to interpersonal relationships and conflicts, to food frustrations.

How to Look after Yourself in Space

A lot of the measures NASA is developing to protect astronauts are not necessarily exclusive to space. They prescribe practices like daily exercise, vitamin supplements, and a clean living environment. Exercise is the key. Working out can activate the immune system cells from our peripheral tissues into our bloodstream. With more of these cells around, there is a higher chance of defeating illnesses. On top of that, exercise maintains muscle and bone, which are at risk in space.

On the mental side of things, astronauts get downtime outside their working schedules. Exercise can offer a mental break, but astronauts also have designated times to contact their families back home. Levasseur says they can bring books or hard drives with their favorite shows up to the station.

"It's a myriad of problems that astronauts have to deal with." "So we have to counteract the diseases with a myriad of solutions."

Why Fly People into Space?

We argue that the goals of a human spaceflight program should satisfy four principal criteria:

1. They should only be accomplishable by humans.
2. They should have benefits that exceed the opportunity costs.
3. They should be worth the risk to, and possible loss of, human life.
4. *Secondary objectives* have benefits that accrue from human presence in space but do not by themselves justify the cost or the risk. These objectives include science, economic development and jobs, technology development, and education.

If human beings are to travel in space for long distances and duration, then it is ethically imperative to understand the biomedical implications for those travelers of prolonged exposure to space and planetary environments.

Importance of Medicine in Space

Space medicine focuses specifically on prevention, acute care, emergency medicine, and hyper/hypobaric medicine in order to ensure medical care of astronauts and spaceflight participants. Medicines are chemicals or compounds used to prevent, halt, or cure disease; to relieve symptoms; or help in the diagnosis of illnesses. These days, medicines come from a variety of sources, and advances in medicine have enabled doctors to cure many diseases and save lives.

Other Examples

Microorganisms include viruses, bacteria, and other infectious pathogens. These pathogens harm living organisms by attacking immune systems, resulting in an array of negative health consequences. Microorganisms are naturally occurring but are transmitted as a result of human interactions.

Extreme weather conditions and natural disasters pose physical health risks to humans and other living organisms. Destruction to shelters, interruptions to food and water supplies, and physical danger from weather events are a few of the ways that these conditions create a hazardous environment.

Bibliography

1. Thirsk R. Physicians as astronauts. Mcgill J Med. 2011;13(2):69. (Wikipedia, The free encyclopedia:Doctors in space)
2. Deiss H. Astronaut requirements. Mcgill J Med. 2020. (The Space Doctors: BMJ 383:p 2726, 2023).
3. Report of the Inter-Agency Meeting on Outer Space Activities on its forthy-third session and its seventhiet open session. (New York, 28 and 31 October 2024, and Vienna, 20 November 2024).
4. Berk LJ, Muret-Wagstaff SL, et al. Inspiring careers in STEM and healthcare fields through medical simulation embedded in high school science education. Adv Physiol Educ. 2014;38(3):210–5.
5. Cohen AJ, Brauer M, et al. Estimates and 25 year trends of the global burden of disease attributable to ambient air pollution: an analysis of data from the global burden of diseases study 2025. Lancet. 2017;389:1907.
6. Vest CE. The effects of the space enviroment on spacecraft surfaces, Johns Hopkinns APL. Tech Dig. 1991;12)(1). (NASA, "Space and its Exploration: How Space is Explored". NASA.gov. Archived from the original on 2 July 2009.)
7. Lea R. The chemical contamination of the International Space Station is out of this world (and not in a good way), August 9, 2023.
8. Letaw JR, Silberberg R, Tsao CH. Radiation hazards on space missions outside the magnetosphere. Adv Space Res. 1989;9:10.
9. McGarrigle P. Hazardous biological activities in outer space. 18 Akron L Rev. 1985;103:4.
10. Lee RJ, Tao Z, et al. Unseen risk: mapping contamination hazards to enhance risk perception in Galena Park, Texas. Clim Risk Manage. 2023;41:100532.
11. Tak S, Calvert G. The estimated national burden of physical ergonomic hazards among US workers. Am J Ind Med. 2024;67(8)
12. Columbus laboratory, European Space Agency (ESA), Jan 10, 2009. March 8, 2009.

26 Space, Science, and Medicines: Innovation and Scientific Advice

Discussion and early interaction are essential with public or private entities engaged in the development of new methods and technologies applied to pharmaceutical products, such as advanced therapies, borderline products, and nanotechnologies.

Space exploration and the innovation it entails are vital drivers for opening up new domains in space science and technology. They trigger new partnerships and develop capabilities that create new opportunities for addressing global challenges. Space exploration also motivates young people to pursue education and careers in science, technology, engineering, and mathematics (the STEM disciplines).

Though the precise nature of future benefits from space exploration is not easily forecast, current trends suggest that significant advantage may lie in areas such as new materials, health and medicine, transport, and computer technology. As the benefits of exploration and innovation become better known, increasingly more countries and non-governmental entities are showing interest in engaging in these fields.

In the past, developments in technology and spin-offs from space research have consistently found their way into communities worldwide, sometimes with lifesaving benefits. The ISS supports experiments that have led to discoveries and inventions in communication for health procedures and robotic surgery. Satellite-enabled Earth observation systems to monitor natural disasters, climates, and crops often facilitate early warnings for threats and mitigation strategies. Space exploration has always been relevant to everyone no matter the discipline or interest.

Drug Development Beyond Earth

The idea of conducting drug development in space is a rapidly evolving field with the potential to change the landscape of health care and pharmaceuticals. While space research has faced its share of challenges and costs, the development is promising.

R. G. Russo, *Navigating Health Challenges in Space*,
https://doi.org/10.1007/978-3-032-03620-9_26

The microgravity in space provides a unique, unparalleled setting for drug development experiments. This environment has the potential to revolutionize medical research by offering new opportunities and opening up avenues for discovering novel drugs, clarifying their mechanisms of action, and improving drug delivery systems. Drugs development in space holds great new promise for advancing medical science and improving healthcare outcomes.

The Promise of Space-Based Drug Development [1]

The goal is to make drug delivery and manufacturing safer and more convenient. Microgravity offers a superior setting for growing high-quality crystals, as it reduces sedimentation, allows precise temperature control, and results in smaller, more uniform particles.

NASA is planning to expand its space-based resources in collaboration with private industries, potentially creating a bustling space economy. Collaboration and knowledge-sharing among pharmaceutical companies are essential to foster these opportunities.

Improving Drug Development in Space [2]

The unique nature of microgravity in space offers opportunities for drug discovery and development that cannot be replicated on Earth. From the production of high-quality protein crystals to the identification and validation of new drug targets, to microarray analyses of transcripts attenuated by microgravity, numerous examples demonstrate the benefit of exploiting the space environment. Studies conducted on space shuttle missions, the ISS, and other spacecrafts have directly benefited drug development programs such as those against reducing bone and muscle loss or boosting bone formation.

While the primary aim of space-based drug development is to maintain astronauts' health and equip humanity for longer ventures into space, the potential benefits extend to people on Earth. Research in space is slow, and the costs are high, with all projects currently subsidized through NASA. In space, we are not only exploring the cosmos but are also discovering new frontiers in pharmaceuticals, revolutionizing health care, and improving the quality of life for people on Earth.

Beyond Space: Impact on Medicine [3]

The integration of pharmaceutical research with space science presents an exciting avenue for exploration: as expected, the absence of gravity offers advances in drug development.

Drug development in space should greatly advance medicine, improving healthcare outcomes. In the absence of gravity, scientists can study the behavior of drugs and their interactions with biological systems in ways not possible on Earth.

From the complexities of human biology to the unpredictability of diseases, researchers constantly battle with variables that can alter the course of a study. In microgravity, cells and molecules behave differently. This unique environment can provide insights into diseases and their treatments that are impossible to achieve here on Earth. As a conclusive remark, microgravity offers unique advantages when investigating cells and proteins, but collaboration and a strong ethical framework are needed to fully realize its benefits.

Cardiac Discoveries in Orbit

Cancer takes a heavy toll on humanity, but the burden of cardiovascular disease is even greater, accounting for 20 million fatalities each year globally. In 2018, NASA launched a study to explore the development of heart cells, cardiomyocytes, in the unique environment of space. This concluded that specific sets of stem cells are more likely to develop into cardiomyocytes. Building on this insight, another study concluded that these heart cells not only developed more frequently in space but could also be used to treat heart problems caused by long spaceflights or heart diseases on Earth.

Governance in Space-Based Research [4]

As we venture deeper into the cosmos, our guiding principles in space research must be firmly based on ethics, transparency, and collaboration. Disease and drug research demand a robust and adaptive governance framework, especially when conducted in the unique environment of space. The concept of a space research "consortium" offers promises. More than just a collaborative platform, it would set universal ethical standards for research in space. That would ensure that discoveries made beyond our atmosphere transcend geopolitical boundaries and are accessible and beneficial to all.

A New Chapter in Laboratory Research

Space presents unique tools for clinical laboratory research. As we decode the mysteries of cells in microgravity and grasp the potential of proteins crystallized in space, we are writing a new chapter in medical science.

Space: The Next Frontier in Drug Development

Space exploration and the innovation it entails are essential drivers opening up new domains in space science and technology. They trigger new partnerships and develop capabilities that create fresh opportunities for addressing global challenges. Space exploration also motivates young people to pursue education and careers in science, technology, engineering, and mathematics (the STEM disciplines).

Though the precise nature of future benefits from space exploration is not easily defined, current trends suggest that significant advantage may lie in areas such as new materials, health and medicine, transport, and computer technology. As the benefits of space exploration and innovation become better known, more countries and non-governmental entities are becoming increasingly interested in exploration and innovation. Over the next decade, NASA is planning to expand its space-based resources in collaboration with private industries to build infrastructure that could 1 day lead to improvements in drug delivery and manufacturing and give a boost to fields like organ transplantation.

The microgravity in orbit is a far better setting for growing high-quality crystals because there is less sedimentation, molecules move more slowly, and scientists can precisely control the temperatures in experiments.

> One thing is for certain, space has always been the inspiration for the next generation of innovators and creative thinkers. Architects of new ideas in this era will inspire many more. Ingenuity must also come from academic and research institutions building a new space-ready generation through innovative curricula, scholarships, and research opportunities for key fields at all levels. Most of all, engaging participation is a responsibility anyone can take by steering the conversation and gathering ideas on how we can make this era one of positive benefit for all, while making opportunities inclusive to all. (Seetha Raghavan, Aug. 4, 2021)

Hopefully, there will be space modules dedicated to biomedical research. Since the cost and availability will be more accessible, we can hope to do the iterative type of experiments and plan the broader applications we have not seen yet.

Medical and Biomedical Research [5]

The effects of microgravity on biological, chemical, and physical processes that can be observed in space-based laboratories enable both public and private actors to extend their understanding of a broad range of diseases and to improve medicine and treatments, or develop new ones, for humans on Earth. For example, the absence of sedimentation allows cells to grow faster and crystals to form more and have significantly better quality, making space an ideal environment to speed up research and testing for the pharmaceutical and biomedical industry. The formation of protein crystals is a key to the pharmaceutical industry in the fight against certain diseases (Alzheimer's, diabetes, Parkinson's, etc.).

OneHealth: Environmental Risks and Recovery [6]

OneHealth is an interdisciplinary approach taking account of the links between humans, animals, and the environment. Increasing environmental devastation, habitat degradation (through deforestation, desertification, and urban sprawl), and accelerating effects of global warming are all leading to increasing interactions among domestic animals, wildlife, and humans. More than 75% of emerging infectious diseases in humans are zoonotic, i.e., shared between animals and humans, and the majority of these can be traced to wild animal reservoirs, such as the West Nile virus, SARS, Ebola or monkeypox viruses. Global climate change and warming are expected to have an impact on airborne, food-borne, rodent-borne, vector-borne, and waterborne diseases.

Environmental pollution involves risks for human health. Satellites can monitor air pollution and help identify links with non-communicable diseases such as asthma or coronary artery diseases. Pollens, asbestos, volcanic ash, and wildfire smoke are all extremely important.

Emergency and Disaster Management

Disaster management aims to reduce or avoid the potential losses from hazards, providing prompt and appropriate assistance for disaster victims and working for rapid and effective recovery of normal conditions.

Scientists around the world are studying the effects of microgravity and cosmic radiation via the "off-Earth" ISS laboratory platform. Over 300 medical experiments have been completed to date, with the goal of extending the knowledge gained for the benefit of humanity.

As scientists work to build the pharmaceutical tools for the future of space travel, data sharing and patient privacy are becoming key concerns. Information about each patient usually includes clinical, medicinal, biological, and chemical research as well as fundamentals and practical applications.

Coming back to Earth: Whenever humans invest in space research, unexpected new technologies on Earth tend to follow.

Bibliography

1. Shi En Kim. Pharma goes to space. C&En. 2022;100(40):18.
2. Bilodeau K. Space—the next frontier in drug development. International Space Station; 2023.
3. Seoane-Viaño I, Ong JJ, et al. To infinity and beyond: strategies for fabricating medicines in outer space. Int J Pharm. 2022;4:100121.
4. del Viterale FC. Global governance of the space system: A multilevel governance analysis. Systems. 2024;12(9):318.
5. Donoviel D. Why space is the next frontier to accelerate medical research. Spacenews; 2024.
6. Prata G, Ribeiro AI, Rocha-Santos T. Chapter 1: One Health, pp 1–31, 2022.

27 NASA + ISS

What Is a NASA Spin-Off?

"Spin-offs" are the scientific and technological discoveries and applications that offer unprecedented opportunities to improve human health and save lives. A NASA spin-off is a technology, a product, or a discovery originally developed for space missions but then found practical and commercial applications in other fields. The aeronautical and space objectives achieved by the space agency in the past decades, in fact, required progress that involved practically every scientific and technological discipline: it soon became clear that these innovations formed a precious resource, ready to be used by public organizations and private companies to develop new products or improve existing ones on the market.

How Has NASA Helped Medicine?

NASA is a US government agency responsible for science and technology related to air and space. The Space Age started in 1957 with the launch of the Soviet satellite Sputnik, and NASA was established in 1958.

NASA has around 30 flight surgeons, and two are usually assigned to each mission—a primary and a deputy. Most flight surgeons never leave the planet though one has completed four missions and has been assigned a fifth. However, the general feeling is that "There's no requirement for a doctor to be in space."

NASA explores the unknown in air and space, innovates for the benefit of humanity, and inspires the world through discovery. The Exploration Medical Capability (ExMC) element of NASA's Human Research Program (HRP) promotes human health and performance in space by advancing the design of "medical systems" and risk-informed decision-making for exploration beyond low Earth orbit.

NASA's contributions to medical breakthroughs are extensive, including advances in magnetic resonance imaging (MRI), a medical imaging technique that

R. G. Russo, *Navigating Health Challenges in Space*,
https://doi.org/10.1007/978-3-032-03620-9_27

uses a magnetic field and computer-generated radio waves to create detailed images of the organs and tissues in your body [1], and development of the left ventricular assist device (LVAD) for heart patients, showcasing the agency's role beyond space exploration. (7 March 2024).

Medical and Clinical Experience

The Human Health and Performance (HH&P) Medical and Clinical teams provide specialty medical care as well as occupational and industrial health services to the civil servant, contractor, and astronaut populations at Johnson Space Center (JSC) and its satellite facilities. This includes occupational, aerospace, dive, travel, and preventive medicines to ensure the health, productivity and wellness of JSC team members through the recognition, evaluation, control, and elimination of illness and disease associated with the work environment.

Providing a Variety of Services

Functional areas include the Occupational Medicine Clinic, the Flight Medicine Clinic, the Test Subject Screening (TSS) Group, the Human Test Support Group, the Lifetime Surveillance of Astronaut Health (LSAH) Group, the Clinical Laboratories, the Employee Assistance Program, and Occupational Health Services. The Occupational Medicine Program provides Occupational Safety and Health Administration (OSHA) job-related physical examinations and is responsible for the evaluation and treatment of work-related injuries or illnesses. The Flight Medicine Program provides occupational and aerospace medicine to active astronauts and Aircraft Operations Division (AOD) flight crew members.

Protecting the JSC Community

The JSC Clinical Laboratories are comprised of two labs: Clinical and Microbiology. The Clinical Laboratory is responsible for the analysis of human blood, urine, or other fluids in the areas of hematology, chemistry, immunology, and urinalysis. The Microbiology Laboratory is responsible for the identification of bacteria, fungi, and parasites in human specimens. Another important part of protecting the JSC community is through the Occupational Health Services this team provides. These services include:

- Industrial hygiene
- Environmental surveillance
- Radiological health and safety
- Hazard communication
- Health training

- Hearing conservation
- Chemical management
- Ergonomics
- Emergency operations support to all of JSC

What Is NASA's Mission for Medical Research in Space? [2]

NASA's Human Research Program aims to enable space exploration beyond low Earth orbit by reducing the risks to human health and performance through a focused program of basic, applied, and operational research leading to the development and delivery of measures for human health, performance, and habitability. (12 Feb. 2024).

The Human Research Program

The Program involves research to develop methods to protect the health and performance of astronauts in space. HRP is using ground research facilities, the International Space Station (ISS), and similar environments to foster cutting-edge science.

What Are NASA's Medical Discoveries? [3]

NASA's innovations have led to a variety of medical devices and procedures, including ear thermometers, insulin pumps, artificial heart defibrillators, and improvements in mammography, demonstrating the broad impact of space technology on healthcare.

Aerospace medicine focuses on research, clinical care, and operational support for the health, safety, and performance of crew members and passengers in air and space vehicles, together with the backup staff who help operate these vehicles.

A collaboration initiative between the Italian Space Agency (ASI, Agenzia Spaziale Italiana) and the Department of Public Health of the Federico II University of Naples is under way as part of the Level II Master's Degree in Aerospace Medicine. This degree, required by the Air Force Medical Corps as part of the framework agreement between the Air Force and this university, covers a training gap of approximately 10 years since the closure of the only Italian specialization school in aeronautical and space medicine.

There are many medical advances that came at least in part from NASA:

- Digital imaging breast biopsy system, developed from Hubble Space Telescope technology
- Tiny transmitters to monitor the fetus inside the womb
- Laser angioplasty, using fiber-optic catheters

Moreover, other precious tools are as follows:

- Forceps with fiber optics that enable doctors to measure the pressure applied to a baby's head during delivery
- Cool suit to lower body temperature in treatment of various conditions
- Voice-controlled wheelchairs
- Light-emitting diodes (LED) for help in brain cancer surgery
- Foam used to insulate space shuttle external tanks for less expensive and better molds for artificial arms and legs
- Programmable pacemakers
- Tools for cataract surgery

EASA and Medicine

The provisions regarding the *medical* certification of aircrew and air traffic controllers (ATCOs) are set out in several European Regulations. This certification ensures that pilots are physically and mentally fit to operate an aircraft safely and is designed to assess various aspects of an individual's health, including cardiovascular fitness, vision, hearing, and overall well-being.

General Information About NASA

NASA, the US space agency, showcases its activity, all aimed at keeping the United States at the forefront of the exploration and use of space, and presents the report for the 2023 fiscal year, in which it clearly illustrates how it spends its budget of $25.4 billion, 0.5% of the US budget.

Four hundred pages full of numbers and information have been written that illustrate important results, given that the funding that the State provides to NASA causes an economic output in American society of 75.6 billion dollars.

NASA pays a lot of attention to distributing funding and resources across the national territory; it is easy to see from the report that 45 states have recorded an impact of more than ten million dollars, and also that eight of these, the most interested in space-related technologies, have exceeded the threshold of one billion dollars.

How Does One Become an Aerospace Doctor? [4]

Minimum qualifications are as follows: A degree in medicine (D.O. or M.D.) with a clinical board certification or board eligibility is required. Moreover, one must have at least 1 year of clinical experience in aerospace medicine (military or civilian).

Only five EM-trained American physician-astronauts have flown in space. At least 1 year of clinical experience in aerospace medicine (military or civilian) is required. Operational flight surgeon experience or an accredited Aerospace Medicine residency, and knowledge of current operational activities involved in manned spaceflight are highly desired. A knowledge of international protocol and customs and the ability to speak a foreign language is ideal. Experience with computer systems is important.

Emergency medicine [5] training offers many features that make emergency physicians the best potential medical providers for human spaceflight: competence for rapid assessment and treatment, resuscitation skills, ability to perform a wide range of advanced procedures, exposure to diverse pathologies, and creativity in resource-limited settings. Given that the models for extended missions are based in part on experience from extreme environments and remote settings, training in "wilderness medicine" will complement the skills necessary for future space endeavors.

In addition, all emergency medicine graduates must demonstrate competence with bedside ultrasound, which is the main imaging modality in space because of its portability and the fact that images are not affected by space radiation. The ultrasound equipment in the ISS has proved useful for examining the musculoskeletal system, conducting FAST examinations, renal and bladder surveys, ocular checks, and even nitrogen bubbles in case of decompression sickness.

The program will face many challenges in clinical management as space exploration progresses to longer missions, interplanetary travel, and extraterrestrial colonization.

It is now suggested that a physician should be included in every crew on long-duration spaceflights in order to ensure mission safety. Physician-astronauts will be a valuable asset to these missions because of their adaptability, working as scientists, engineers, managers, and clinicians who can manage space emergencies and provide long-term crew health monitoring. Their training and experience make emergency physicians the ideal specialists for providing these services.

"The knowledge, skills, and attitudes of a clinician are valuable but not sufficient, of course, to be an astronaut: a burning passion for space exploration is also required."

A physician is in fact rarely included as part of a spaceflight crew. In the past 50 years, more than 400 people have flown in space, but only 26 have been American physicians. Surprisingly, few serious medical emergencies in space have been reported. Documented conditions have included urosepsis, renal colic, cellulitis, pneumonitis, and cardiac arrhythmia, but only 17 of these nonfatal severe medical events were reported during spaceflight between 1961 and 1999.

Astronauts are additionally faced with the challenges of medical resuscitation in a microgravity environment, where usual physiologic conditions and advanced life support protocols may not even be relevant.

Basic Course in Aeronautical and Space Medicine

The course in aeronautical and space medicine provides an in-depth study of all the issues relating to pathological and physiological changes in flight and aerospace flight conditions. From a purely professional point of view, it enhances the certification implications, and the good prevalence of students from the fields of occupational medicine and forensic medicine is worth noting. The specialization course is recognized among the training activities that can contribute to obtaining the authorization of Aeromedical Examiner (AME) and offers cultural and professional enrichment for all the doctors involved in healthcare services in the aeronautical sector (airport health services, etc.).

Course Overview

The Basic Course in Aviation Medicine (BCAM) provides fundamental theoretical and practical training in physiology, psychology, and clinical practice as related to aerospace medicine. It meets the educational requirements to apply to become an Aeromedical Examiner with Class 2 medical certification privileges.

Education is a fundamental part of NASA's work to fulfill its vision to reach new heights and reveal the unknown so that what we do and learn will benefit all mankind. NASA's Office of Education collaborates with NASA experts to connect learners and educators to exciting NASA missions through dynamic STEM content. These efforts produce innovative and informative educational materials that arouse interest in science, technology, engineering, and mathematics. NASA makes these resources available in several ways: download and view education resources from NASA's website: http://www.nasa.gov/education. Visit a NASA Educator Resource Center (ERC): http://www.nasa.gov/education/ercn Educational Multimedia.

NASA offers educational multimedia such as the Do-It-Yourself Podcast activity for students, video learning clips, webcasts, monthly educational programming on NASA TV, and image galleries.

For more information, visit: Do-It-Yourself Podcast http://www.nasa.gov/audience/foreducators//diypodcast/index.html, http://blogs.nasa.gov/diyPodcastBlog/, NASA and clips (video learning clips) http://www.nasa.gov/audience/foreducators/nasaeclips/index.html, NASA's Digital Learning Network (webcasts) http://www.nasa.gov/offices/education/programs/national/dln/index.html, and image galleries.

Key Benefits

- World-class learning delivered at a location recognized by the Civil Aviation Authority for training AMEs
- Teaching by internationally recognized experts in the field
- Meets the training requirements of EASA Part Med Subpart D (but is not currently an EASA-accredited course)

UC (Univ. Calif.) Mission

The mission of the UC "Space Health Program" is to address the complexities of optimizing human health and medical management in space and return the knowledge to Earth. Previous human space expeditions have deepened our understanding of physical and behavioral adaptations to the spaceflight environment and the measures to address them.

Research and development applied to the Space Health Challenge also deepen our knowledge of physiologic responses on Earth, providing models of extreme environments and accelerated aging and technological advances that enhance terrestrial healthcare. The advanced technology requirements for human space exploration on increasingly longer missions contribute significantly to remote self-healthcare and medical management on Earth.

In collaboration with NASA and industrial space health experts, including astronauts, clinicians, scientists, engineers, and health tech innovators, we seek to understand how space-related science and technology can accelerate medical innovation to optimize crew health and safety and provide health benefits universally.

The UC has a long history of pioneering biomedical research and vision for advancing health, worldwide and galaxy-wide! More than 150 workers in the UC are engaged in research and innovation related to advancing human space exploration. UC Space Health is a cross-campus and cross-discipline network that provides opportunities to share resources and knowledge and to collaborate in research and funding opportunities.

Why Is Space Exploration Important in Medicine?

Science experiments on astronauts in space improve our understanding of medical conditions on Earth. This research has produced findings that can help people suffering from cardiovascular disorders, type 2 diabetes, osteoporosis, and balance problems.

Medical research on astronauts on the ISS accumulates a large range of data, collected before, during, and after the astronauts' time in space. The same data are sometimes employed for different research projects.

Microgravity and Stress

In space, since gravity no longer pulls blood and fluids to our legs, it all shifts headward. The body's response is to lower its blood volume to restore homeostasis. One explanation for the downregulation of immune-related gene activity is that the number of circulating red and white blood cells must also decrease—as there are too many of these cells in a smaller amount of blood. This in turn can affect how the immune system defends itself.

Stress is an additional factor. Space missions have high stakes, from launch to landing. Experiencing constant stress in a physically and socially isolated environment, far from family and friends, where many of our natural rhythms are disrupted, places heavy burdens on the body.

ISS

Problems on the International Space Station: An Air Leak Detected—What's Happening

Past studies had shown that bacteria grown on the ISS have a frequency of mutations tens of times higher than the same ones present on Earth in genes linked to drug resistance. The 13 strains analyzed, as explained by the scientists in the abstract of the new study, show "resistance mechanisms that classify them within the ESKAPE [6] group of pathogens. The acronym ESKAPE, coined by the American Society of Infectious Diseases in 2008, refers to the following bacteria that present high rates of multidrug resistance: *Enterococci*, *S. aureus*, *K. pneumoniae*, *A. baumannii*, *P. aeruginosa*, Enterobacteriaceae."

Genomic analysis also highlighted the presence of genes not observed in terrestrial counterparts. These substantial differences with the strains cultivated on Earth, according to experts, would be linked precisely to the selection determined by the extreme environment in which they live, "bombarded" with solar and cosmic radiation, in microgravity and with high concentrations of CO_2.

The surviving colonies facilitated by the mutations have adapted well to various environments of the ISS; bacteria, in fact, can be found in abundance in various places of the orbiting laboratory. Furthermore, the isolated strains coexisted with other microorganisms, some of which could probably survive thanks to the presence of *E. bugandensis*.

"The implications of these findings are twofold. First, they shed light on the behavior, adaptation and evolution of microbes in extreme and isolated environments. Second, they highlight the need for robust preventative measures, ensuring the health and safety of astrted with potential pathogenic threats," Venkateswaran and colleagues explained in the study.

International Space Station

The Italian astronaut and aviator Samantha Cristoforetti, the first European woman at the head of the International Space Station, spent 369 days whizzing at 27,000 km/h in orbit around the Earth, about 400 km above our heads, with two missions.

As a result, the space station has aroused curiosity. After departing from the Baikonur Cosmodrome (Kazakhstan) at 10.01 pm Italian time together with colleagues Anton Shkaplerov and Terry Virts, Samantha Cristoforetti completed a

6-hour flight aboard the Russian Soyuz, anchoring and entering the station at 6.00 am. She lived for 6 months on the ISS, orbiting the Earth, and took part in around 200 scientific experiments designed by the Italian Space Agency (ASI).

More recently, in October 2024, on the occasion of the seventy-fifth edition of IAC 2024, the astronaut presented the ambitious project for private space stations on which she is working on.

The ISS Today

The ISS is the result of cooperation between five space agencies: NASA (United States), Roscosmos (Russia), European Space Agency, JAXA (Japan), and CSA (Canada). It consists of numerous modules linked together. It has a habitable volume of approximately 1200 cubic meters. It typically hosts six or seven astronauts, but it can reach up to eleven. The International Space Station is essentially a large orbiting laboratory. Many do not hesitate to define it as the most complex engineering work ever built by man.

What Is the ISS For?

To start with, the ISS is first of all a large scientific research laboratory, where it is possible to conduct long-term experiments on the effects of almost total weightlessness, tests that could not have been carried out on the space shuttles that remain only short times in orbit. Being in an environment different from that of Earth, the ISS offers optimal unusual conditions for experiments on biotechnology, chemistry, medicine, physics, meteorology, materials science, and astronomy. The studies focus for the most part on the human organism and its adaptation to weightlessness: this information is essential to understand how to develop new systems and technologies for future space exploration. By studying reactions in an environment different from that on Earth, we can also discover many things about the human body that can be useful for our everyday lives, thanks to medical research.

The International Space Station led to breakthroughs in human health on Earth. Major new avenues are as follows:

1. *A new way to look at Alzheimer's*

Researchers from around the world use the space station to address complex human health problems on Earth. Studies examining the formation of amyloids started after evidence suggested similarities between their formation and protein crystallization, a process already extensively studied in space.

2. *Breathing easier in space and on Earth*

Research aimed at protecting the lungs and airways of astronauts is helping people with asthma and other breathing issues on Earth. Experiments conducted aboard the ISS were collaborated on by medical technologists at Sweden's Karolinska Institute, who developed an instrument that measures exhaled air for nitric oxide.

3. *Toward better cancer treatments*

The ISS National Laboratory advanced cancer research through studies that cultivate stem cells for possible therapeutic applications, crystallize proteins for improved drug discovery and delivery, and tested therapies to find those that work better or have fewer side effects.

4. *Insight into the immune system*

Astronauts experience impaired immune function in space. Space offers a unique advantage to studying this aging-related immune suppression since conditions that normally develop over decades happen quickly there. A growing body of research took advantage of this fact.

The Leukin-2 experiment demonstrated that microgravity has a fundamental effect on activation of T cells, which are involved in turning on and off the immune system. Increased understanding of how stress affects the immune system could help people on Earth as well as support development of drugs to counter immune system changes for long-duration spaceflight.

5. *Protecting muscles and bones*

The ISS National Lab aimed to maintain muscle and bone, protect astronauts against bone and muscle loss in microgravity, and even stimulate growth. Results suggested that this strategy could prevent and treat such loss in people on Earth as well. There is still a fair amount of work to be done to pursue this strategy for treating muscle loss.

By providing a platform for long-term microgravity research, the space station fosters breakthroughs that yielded direct benefits to people on Earth.

Who Does the ISS Belong to?

Ownership of the ISS is regulated by a series of treaties signed by 11 of the European countries that are part of ESA, along with the US NASA, and the Russian and Japanese space agencies. These agreements establish the methods of access to various modules and the time each space agency can dedicate for its research. The plans of the 14 nations that contributed to building it originally envisaged that the ISS would remain in operation until at least 2020, but for some time, there has been talk of extending the missions so that it can be inhabited until 2028. In all, the expected cost for the 30 years of management, including construction, was estimated at around 100 billion euro. The ISS orbits between 350 and 400 kilometers above sea level, about thirty times the altitude at which a commercial airliner normally travels.

How the ISS Is Made

The ISS offers approximately 800 cubic meters of habitable space and is 100 meters long. It was built piece by piece by joining together large cylindrical modules. Some were created by the Russian space agency, such as Zvezda (Stella) and Zarya (Alba), and others by the United States and Europe, with a particular commitment on the part of the Italian Space Agency. At the center of the Station, there is a long axis from which the 16 main solar panels open, which provide energy to the ISS.

There are usually six astronauts on board the Station. During long-term missions, like that of Samantha Cristoforetti, they stay on board for about 6 months to understand the effects of weightlessness on the human organism and explore ways to counteract its consequences.

Safety on board has priority over other jobs, and hundreds of hours of work are required to maintain the Station every week.

Life on Board

Living in an environment constantly subjected to the force of gravity is one of the things our body does best. In the almost total absence of weight, our body finds itself in difficulty: in the first few days, body fluids tend to rise toward the head and upper limbs, which explains why astronauts who have just arrived on the ISS usually appear purple and visibly swollen. Many tenants of the Station also have stomach problems because digestion is a process that takes place *by definition* from top to bottom.

A long time in an environment where you weigh very little is also harmful to the muscular and skeletal systems. It is estimated that for every month an astronaut stays on the ISS, he loses up to 2% of his skeletal mass. This is why astronauts on the Station must do gymnastics every day to maintain muscle tone and reinforce bones.

Breathing and Washing

The ISS environment is completely sealed, so the air circulating inside has to be constantly purified. The aeration system introduces oxygen from the Station's large tanks, which are filled periodically thanks to the transport shuttles that bring supplies of all kinds into orbit. For hygiene reasons and to prevent humidity from damaging on-board devices, the purification system makes the air very dry and this means that astronauts' senses such as taste and smell are constantly in difficulty.

In an environment full of electronic equipment, as little water as possible must circulate. Bags with a straw are used for drinking, so no drops remain suspended around the modules. Washing takes time. Even going to the bathroom is not easy, but things have improved a lot since a toilet equipped with a suction system was installed.

Day and Night

The ISS takes about an hour and a half to complete a complete revolution round the Earth: it therefore completes 16 orbits in one Earth day. The Station travels quickly around the planet, but you get the sensation of being still, while something slowly turns the terrestrial globe. Astronauts sleep in the dark, with earplugs and masks.

Tolerance of Acceleration

During takeoff and during the outward journey, the human body gradually reaches a final speed of about 25,000–27,000 km/h in about 10 minutes. However, this enormous acceleration has no effect on astronauts because it feels gradual. The same holds during the deceleration phase during the return [7].

The Challenge of Space Medicine to Make Us Feel Better on Earth

This medical branch has experimented, tested, and approved therapeutic solutions especially for astronauts; has accelerated aging; and determined changes that usually occur in 10–20 years of life on Earth, with deleterious effects on DNA, heart, metabolism, and eyes. But from the disorders that affect astronauts upon their return from space, there are also indications for preventing and treating aging diseases on Earth. While this may seem harmful, it actually represents a unique opportunity for science. In fact, the rapid aging simulation can be inverted and analyzed in depth. Furthermore, in microgravity, the brain must adapt its information processing mechanisms, offering precious traces on human cognitive abilities and paving the way for next generations of explorers, including civilian ones.

Artificial intelligence programs are being studied and refined, capable of diagnosing pathologies even before the symptoms appear, as well as liquid biopsies which, with a blood sample, can recognize the "signs" of different types of tumors. All of this extends to the most advanced telemedicine systems and the new frontiers of pharmacogenomics. All innovations were designed for astronauts, but they can also help us "terrestrials."

What Is the Future of the Space Station?

When the ISS retires, the plan is to detach the Axiom module(s), allowing them to operate as an independent space station that NASA and other customers will be able to use for a fee (Aug. 30, 2024).

Does the Future of Medicine Lie in Space?

NASA suggests that its astronauts could multitask, using their time in orbit to pursue a cure for cancer or the many other illnesses afflicting humankind.

New Possibility for the ISS?

NASA is not convinced of its replacements.

Probably by the end of 2030, the ISS will be decommissioned. In its place, new private space stations will take over. The first is expected to be Axiom Space. Orbital Reef by Jeff Bezos' Blue Origin and Sierra Space and the Starlab by Voyager Space and Airbus are also already being planned for the following years.

The idea of having a human stay in space has fueled dreams of colonies on other planets, of new life forms, and of a deeper understanding of the cosmos. This commitment, made visible with the International Space Station (ISS), has allowed scientific research to progress in extreme contexts, such as that of microgravity.

We are not just talking about sending astronauts into orbit, but about ensuring that human permanence can be sustainable and continuous. The challenges are numerous, starting from the technical difficulties up to the enormous costs that such missions entail.

The idea of having men and women living and working in space has always been a strategic goal for the world's major space agencies. It is no surprise that the future of orbital exploration is being pondered, especially at a time when private and public technologies are converging.

While companies like SpaceX and Blue Origin are carrying out pioneering innovations, government agencies are evaluating how to integrate these advances with missions of international scientific and security interest.

Bibliography

1. McRobbie DW, Moore EA, et al. MRI from picture to proton. Cambridge University Press; 2007. p. 1. 978-1-139-45719.
2. Abercromby A, Bauman D, et al. NASA's top human system research and technology needs for Mars. Acta Astronautica. 2025;228:931–9.
3. Jaber J, Ong J, et al. NASA's impact on medical innovation: breakthrough technologies from space research. Acta Astronautica. 2024;226(13). National Aeronautics and Space Administration. "NASA Spinoff Frequently Asked Questions." (Accessed March 16, 2011).

4. Specialty guides, The Complete Guide To The Complete Guide To Becoming an Aviation and Space Medicine Doctor, 18 Sep 2024.
5. Hirshon JM, Morris DM. Emergency medicine and the health of the public: the critical role of emergency departments in US Public Health. Emerg Med Clin North Am. 2006;24(4):815–9.
6. La multiresistenza dei batteri ESKAPE (IT. The multi-resistance of ESKAPE bacteria) Assistenza farmaceutica territoriale, ULSS 6, Padova., (IT.Territorial pharmaceutical assistance, ULSS 6, Padua), Last update: 23/02/2022.
7. Glaister DH. Human tolerance to impact acceleration. Injury. 1978;9(3):191–8.

The Balanced System 28

Balance in Space

Out of This World: How Space Messes with the Balance

Essential Points:

- Microgravity disrupts balance systems: In space, the lack of gravitational cues confuses our balance systems, causing astronauts to experience disorientation and space motion sickness (SMS), which can hinder mission tasks.
- Adaptation through neural recalibration: Astronauts adapt to microgravity over time through neuroplasticity, relying more on visual and proprioceptive cues, with training and intentional movement helping them manage SMS symptoms.
- Insights benefit Earthly balance issues: Research on space balance challenges may help develop new therapies for balance disorders on Earth, offering potential benefits for those with vestibular and age-related balance issues.

Astronauts have a hard time with balance in space. SMS (space motion sickness) is disorienting astronauts and at times hindering mission tasks.

We will explore how weightlessness affects balance, the science behind SMS, and what astronauts do to cope with these challenges.

What Is Balance?

Balance refers to our body's ability to maintain its center of mass over its base of support. It is essential for everything we do from walking, running, standing still, and even sitting. Our sense of balance relies on three primary systems:

R. G. Russo, *Navigating Health Challenges in Space*,
https://doi.org/10.1007/978-3-032-03620-9_28

1. *Vestibular system*: Located in your inner ear, this system detects changes in your head position and motion. It helps understand where our body is in space relative to gravity.
2. *Visual system*: The eyes provide crucial information about our surroundings, helping to orient ourself.
3. *Somatosensory system*: This system includes sensors in our muscles and joints that inform the brain about the position of limbs.

All three systems work together to help to maintain balance. But what happens when one of these systems is disrupted?

The Role of the Vestibular System

The vestibular system plays a critical role in balance. It consists of tiny structures called semicircular canals and otolith organs.

- *Semicircular canals*: These three looped tubes filled with fluid detect rotational movements. For example, when we tilt the head or spin around, the fluid moves, triggering hair cells that send signals to the brain about our position.
- *Otolith organs*: These structures respond to gravitational forces (the pull of the Earth's gravity downwards) and linear accelerations (gaining increased speed in a certain direction). They help us to understand if we are upright or lying down.

When on Earth, gravity helps these structures function correctly. However, when in space, things change dramatically.

The Effects of Microgravity on Balance

In the microgravity (less gravity than Earth) environment of space, the normal gravitational cues our body relies on are absent. This absence significantly affects our vestibular system's normal workings, leading to confusion and disorientation.

- *Disruption of signals*: In space, the fluid in the semicircular canals does not move in the same way it does on Earth, which can lead to conflicting signals being sent to your brain. For example, when we turn our head, the fluid may not shift as expected, leaving the brain unsure of our body's position.
- *Otolith dysfunction*: The otolith organs also struggle to detect gravitational cues pulling down on them. Without these signals, astronauts may feel as though they are floating or even spinning, leading to dizziness and nausea.

Space Motion Sickness: The Unwelcome Companion

As a result of these disruptions, many astronauts experience what is commonly known as space motion sickness (SMS). It is essentially the body's response to the new and confusing conditions in space.

Symptoms of Space Motion Sickness

The symptoms of SMS can be quite debilitating and often resemble those of regular motion sickness. Common symptoms include:

- Nausea
- Vomiting
- Dizziness
- Sweating
- General malaise

Approximately 70% of astronauts report experiencing some level of motion sickness during their initial days in space.

In some cases, SMS can even affect an astronaut's ability to perform essential tasks.

Coping with Space Motion Sickness

Adaptation to Zero-Gs

Here is how this adaptation process works:

1. *Neural recalibration*: the brain is incredibly adaptable and has an ability known as neuroplasticity, where it can change the structure and function of nerve connections. Over time, it learns to adjust to the lack of gravity and recalibrates its understanding of spatial orientation, or rather its position in space. This process can take anywhere from a few days to a week.
2. *Compensatory mechanisms*: As astronauts adapt, they start relying more on their visual and somatosensory systems to maintain balance. For instance, they may look at the walls of the spacecraft or focus on their own body movements to help orient rather than trust how their position in relative space feels.

Preventative Measures

To help mitigate the effects of SMS, astronauts undergo extensive training before their missions. Here are some strategies that are often utilized:

- *Preflight training*: Astronauts participate in simulated microgravity experiences, such as parabolic flights (often referred to as the "vomit comet"), to help their

bodies get used to the sensations of weightlessness. These parabolic flights occur when a plane flies an inverted U shape allowing simulated weightlessness to occur by exploiting the laws of physics.
- *Pharmacological interventions*: Some astronauts use medications like anti-nausea drugs to help manage their symptoms during the first few days in space. While these drugs can alleviate some discomfort, they may not work for everyone.
- *Intentional movements*: During the initial adjustment period, astronauts are advised to move slowly and deliberately to minimize the risk of falls or disorientation.

Rehabilitation After Space

Once astronauts return to Earth, they often face a new challenge in the form of readjusting to gravity. After spending extended periods in microgravity, the vestibular system is no longer efficiently adapted to Earth's gravity, leading to balance disorders upon return to the surface that could last up to 30 days.

- *Balance rehabilitation*: Many astronauts undergo rehabilitation to help re-establish their vestibular function. This process includes exercises that challenge their balance and proprioception, allowing them to re-acclimate with the gravitational forces on Earth.

The Broader Implications of Space and Balance

Long-Duration Missions and Future Challenges

As space agencies plan for long-duration missions, such as potential trips to Mars, understanding how microgravity affects balance will be crucial. The insights gained from studying SMS can inform the development of countermeasures for astronauts on these extended journeys.

Potential Countermeasures

1. *Enhanced training protocols*: Space agencies may implement more extensive preflight training to help astronauts acclimate to microgravity more effectively before ever stepping into (or rather floating into) microgravity.
2. *Innovative technologies*: Advances in virtual reality and balance training tools could provide astronauts with additional resources to improve SMS and reduce overall negative effects of vestibular adaptations in space.
3. *Research and collaboration*: Continued research into the physiological effects of space travel can lead to better training regimens, equipment, and strategies to keep astronauts healthy and capable during missions.

Insights for Us Earthlings with Balance Issues

Understanding how the body adapts to microgravity may help develop therapies for people with vestibular disorders or balance issues caused by age or illness. A large handful of products to be used every day were first invented due to the space program, so it is quite likely any developing vestibular technology or rehab tools will trickle down to us normal Earthlings.

Fun Facts About Space and the Human Body

- *Body fluids*: In space, body fluids shift towards the upper body, which can cause a "moon face" appearance among astronauts.
- *Bone density loss*: Without the stress of gravity, astronauts can lose up to 1% of bone density per month, increasing the risk of fractures upon return to Earth, which is further increased by potential balance issues.
- *Height changes*: Astronauts can grow taller in space, with some reports indicating an increase of up to 2 inches due to the expansion of the spine without gravitational pressure.
- *Altered senses*: Many astronauts report changes in taste and smell while in space which can affect how much food they eat, and possibly lead to "space anorexia" in some astronauts.

Final Thoughts: Finding Balance in Space and Beyond

It is clear that the absence of gravity creates a unique set of challenges for astronauts. The vestibular system, which plays a vital role in our ability to stay upright and oriented in Earth's atmosphere, is thrown into disorder under microgravity conditions. This disorder can lead to phenomena like space motion sickness SMS). However, through training, adaptation, and rehabilitation, astronauts can learn to navigate this new environment and remember to always stay curious, keep learning, and embrace the wonders of science, both in space and right here on Earth!

Loss of Balance [2]

Nearly 70% of all astronauts experience impaired balance, locomotion, gaze control, dynamic visual acuity, eye-head coordination, and/or motion sickness in the first 3–4 days of space flight and then again after returning to Earth.

Leaving and returning to Earth's gravity causes "space sickness," with dizziness and loss of balance. By studying how these changes affect the human body, involving the senses, brain, inner ear, and blood pressure, NASA hopes to develop treatments that can be used on Earth and in space to correct balance disorders. Until then, NASA's astronauts must rely on a medication called midodrine (an "anti-dizzy" pill that temporarily raises blood pressure) and/or promethazine to help complete the tasks they need to do to return home safely.

Bibliography

1. Tuttle MH, Kilgore RA, Moore DL, Langley Research Center. Magnetic suspension and balance systems: a comprehensive, annotated bibliography. Washington, D.C.: National Aeronautics and Space Administration, Office of Management, Scientific and Technical Information Program; 1991.
2. Lawson BD, Rupert AH, Braden J, McGrath BJ. The neurovestibular challenges of astronauts and balance patients: some past countermeasures and two alternative approaches to elicitation, assessment and mitigation. Front Syst Neurosci. 2016;10:96.

Sleeping and Eating in Space + Telemedicine

29

The ISS station provides accommodation for each permanent crew member, with two "sleep stations" located in the Russian segment and another four in the Tranquility module. US accommodations are one-person, sound-proofed cabins. A crew member can rest in a sleeping bag, listen to music, use a laptop, and store personal items in a large drawer or in nets attached to the walls. The accommodation includes a reading lamp and a shelf.

Crews visiting the station, who do not have their own assigned accommodation, use a sleeping bag attached to the station walls. Crew quarters are well ventilated because otherwise astronauts could suffocate from the exhaled carbon dioxide that could build up around them.

Sleep Disorders [1]

Both acute and chronic partial sleep loss is frequent in spaceflight due to operational demands and for physiological reasons not yet entirely understood. Astronauts must maintain a high level of cognitive performance during every phase of the mission. Top-tier performance depends on acquiring an adequate quantity of daily sleep and the appropriate sleep quality.

Previous spaceflight experience has shown that astronauts commonly experience sleep deprivation. Additionally, due to the nature of spaceflight, circadian disturbances are present. Together, these two aspects lead to fatigue and errors while performing tasks. Evidence from short- and long-duration missions and other relevant environments suggests that environmental factors (e.g., noise, temperature, vibration, and light) inhibit sleep and impact well-being in space. Thus, for crew members to achieve optimal sleep, they must be provided with a sleep environment that allows them to achieve quality sleep, free of external disruptions.

Spaceflight can disrupt physiological processes that influence sleep patterns in human beings. Astronauts have asynchronized cortisol rhythm patterns, dampened

R. G. Russo, *Navigating Health Challenges in Space*,
https://doi.org/10.1007/978-3-032-03620-9_29

diurnal fluctuations in body temperature, and poor sleep quality. Sleep pattern disruption in astronauts is a form of extrinsic (environmentally caused) circadian rhythm sleep disorder.

Poor sleep quality and quantity can affect the daytime performance and attentiveness of space crew. As a result, improving nighttime sleep has been a topic of NASA-funded research for more than half a century.

Sleep Medicine

The use of sleep aids is widespread among astronauts, with one ten-year study finding that 75% and 78% of ISS and space shuttle crew reported taking such medications while in space. Those who took sleep-inducing medications reported they used them 52% of all nights. NASA allocates 8.5 hours of "downtime" per day for sleep for astronauts aboard the ISS, but their average duration of sleep is only 6 hours.

The following pharmacological and environmental strategies have been investigated in relation to sleep in space [2]:

- Light therapy involving exposure to visible light at varying intensities and wavelengths is a key topic of interest in NASA-funded research. NASA has trialed and implemented rhythmic light panels on the ISS to help entrain the circadian rhythms of astronauts. More advanced light panels that change their output light intensity and wavelengths according to the time of day will be tested soon.
- Melatonin, a naturally occurring hormone secreted by the pineal gland, has shown positive effects in reducing sleep latency in orbit.
- Nonbenzodiazepine sedative-hypnotics (also known as "z drugs") such as zolpidem, zopiclone, and zaleplon are the most commonly dispensed medications on the ISS. Prior research suggests that nonbenzodiazepines may produce less residual impairment than most benzodiazepines. Astronauts frequently take second doses of hypnotic drugs; the shorter duration of action of nonbenzodiazepines may be better suited to middle-of-the-night dosing.
- Benzodiazepines are frequently used medications in space, though less often than nonbenzodiazepine "z-drugs." The longer acting nature of some benzodiazepines used by astronauts, such as temazepam, has been cited as "non-ideal" for spaceflight use due to a high tendency of causing morning impairments.
- Modafinil, a wakefulness drug, is available on the space station to mitigate the deleterious effects of sleep disruption and "optimize performance while fatigued."

The challenges of medicine in space—from aging to effects on "terrestrial" diseases—are artificial intelligence programs capable of diagnosing diseases, liquid biopsies to recognize the "signs" of different types of cancers, digital twins to predict the evolution of diseases, and new telemedicine systems to intervene remotely [3].

The Threats of Space to Health [4]

Space travelers face two main challenges: microgravity and oxidative stress, i.e., an increase in free radicals to a level that compromises the cell's antoxidant capacity and causes DNA damage. In response to all this, the human physiology changes to adapt, and the result is a strong acceleration of aging even by 10–20 years. The continuity of our existence on Earth might face an uncertain future, mainly due to climate changes, the depletion of important raw materials on Earth, collisions with other celestial bodies such as asteroids, the invasion of infectious agents for which we have no defense, and, ultimately, the collapse of the Sun itself—all possible scenarios that make it imperative to explore alternative worlds on which to survive and prepare humanity for life on them.

The lack of gravity in space opens up unique opportunities for drug design and development. Both the discovery and manufacture of drugs in space may become practical, affordable—and, perhaps, commonplace. The drug industry is experiencing a surge in biologic medicines, a class of bulky, biobased pharmaceuticals that includes enzymes, nucleic acids, and antibodies.

A low-gravity environment has several advantages that can accelerate the discovery and preclinical testing of these complex molecules. Behind many diseases, there are proteins such as insulin and interferon. Scientists typically crystallize the protein into an ordered array in the same way that atoms order themselves in a crystal lattice. Protein crystallization is a finicky process that requires tightly controlled conditions. Thanks to microgravity, space-grown protein crystals are often larger and have fewer defects than those prepared on the ground. Producing protein crystals in space to formulate them into a drug suspension may give them more desirable properties than those of their Earth-made counterparts. In such a formulation, the drug must be much more concentrated while maintaining a low enough viscosity to pass through the syringe needle.

Moreover, astronauts have exhibited signs of accelerated bone and muscle loss, a weakened immune system, and altered gene expression after a sojourn in space. Drug developers are already exploiting the biological shifts triggered by space for pharmaceutical testing. Preclinical trials of two osteoporosis drugs are already run, and the researchers found that the treatment preempted space's atrophying effect on muscles.

In other tests of cell growth, researchers have found that stem cells can grow more rapidly in space. Such capabilities would be valuable for treating heart failure as the blood-pumping organ has poor regenerative capabilities. Space might help make stem cell therapy more practical.

A New Era in Space [5]

Researchers prepare for several years for each experiment, revising procedures, doing test runs, miniaturizing hardware, and mercilessly pruning their experiments to reduce the payload weight. Researchers can turn to more accessible albeit imperfect alternatives.

Several companies plan to offer in-space manufacturing services. Given experiments suggesting that stem cells proliferate more quickly in microgravity than on Earth, researchers hypothesize that space would allow for a more efficient and scalable pipeline for supplying stem cells for the clinic.

Oxidative stress derived, in particular, from ionizing radiation, which penetrates thousands of cells at high doses, alters the function of mitochondria, the cell's energy production unit, and consequently the metabolism of carbohydrates and lipids. It also damages DNA, modifies the expression of genes, and alters the length of telomeres. It governs the redistribution of fluids toward the upper part of the body, increasing the risk of thrombosis and vision problems.

Doctors, astronauts, engineers, astronomers, historians, physicists, and ethicists will take a multidisciplinary approach in that unique three-day event.

Experts will discuss the numerous ethical issues that accompany the creation of a new space civilization.

Why Should We Build Space Stations? [6]

There are a variety of reasons for building and operating space stations, including research, industry, exploration, and even tourism. The first space stations were built to study the long-term effects of weightlessness on the human body. After all, if astronauts will ever venture to Mars or other planets, then we must know how prolonged microgravity on the order of months to years will affect their health.

Space stations are a place to do cutting-edge scientific research in an environment that cannot be matched on Earth. For example, gravity alters the way that atoms come together to form crystals. In microgravity, near-perfect crystals can be formed. Such crystals can yield better semiconductors for faster computers or for more efficient drugs to combat diseases. Another effect of gravity is that it causes convection currents to form in flames, which leads to unsteady flames. This makes the study of combustion very difficult. However, in microgravity, simple, steady, slow-moving flames result; these types of flames make it easier to study the combustion process. The resulting information could yield a better understanding of the combustion process and lead to better designs of furnaces or the reduction of air pollution by making combustion more efficient.

Space stations might be used also for space hotels. Here, private companies could ferry tourists from Earth to space hotels for brief visits or extended stays. Even grander extensions of tourism are that space stations could become space ports for expeditions to the planets and stars or even new cities and colonies that could relieve an overpopulated planet.

Skylab: America's First Space Station

The United States placed its first and only space station, called Skylab 1, in orbit in 1973. During the launch, the station was damaged. A critical meteoroid shield and one of the station's two main solar panels were ripped off, and the other solar panel was not fully stretched out. That meant that Skylab had little electrical power, and the internal temperature rose to 126 °F (52 °C).

The Future of Space Stations [7]

NASA is exploring the idea of supporting the development of commercial space stations. The Commercial LEO Development (CLD) program aims to ensure that, once the ISS is retired, there are alternative infrastructures ready to take up its legacy.

The transition will be crucial to maintaining a constant presence in space, in light of the limited resources of these new projects. As technology advances, there may be other solutions to create a coordination center within Earth's orbit.

The ISS will need periodic reboosting because of its position in low Earth orbit. However, there are two places between the Earth and Moon called Lagrange Points L-4 and L-5. At these points, the Earth's gravity and the Moon's gravity are counterbalanced so that an object placed there would not be pulled toward the Earth or Moon. The orbit would be stable and require no boosting. A society called the L5 Society was formed more than 20 years ago to push the idea of placing space stations in orbit at these points.

As we learn more from our experiences on the ISS, we may build larger and better space stations that would enable us to live and work in space, and the dreams of von Braun and the early space scientists may someday become reality.

Eating and Drinking in Space [8]

Most of the food eaten by station crew is frozen, refrigerated, or canned. The menus are studied by the astronauts, with the help of a dietician, before the mission. Because the sense of taste is diminished in orbit, spicy food is a favorite of many crews. Each crew member has individual packets and cooks the food in the on-board galley equipped with two food warmers, a refrigerator, and a hot and cold water dispenser.

30 Telemedicine

Space technologies have left a significant imprint on medical imaging, opening up new horizons for the diagnosis and treatment of various pathologies. NASA has introduced sensors with high-resolution cameras that have proved incredibly useful in medicine, as they have been adapted to improve precision medical imaging techniques. These techniques enable doctors to explore the human body noninvasively, giving detailed views of internal organs and anatomical structures. This makes it possible to diagnose diseases early and plan surgery with maximum precision.

Digital image processing has found a wide range of applications and is used to create and enhance images of organs in the human body for diagnostic purposes, with imaging techniques such as computed tomography and magnetic resonance imaging, for example, in prosthetics such as artificial limbs, synthetic bones, and dental implants. In recent decades, NASA has developed the so-called "intelligent fabrics," capable of monitoring and adapting to the environmental conditions to which astronauts are subjected during space missions. These are also finding applications in medicine as, for example, adjustable elastic bands that constantly monitor vital parameters, providing precious data for optimal management of patients' health.

Finally, the necessary communication between astronauts in space and medical specialists on Earth has led to a large-scale experimentation of telemedicine, which constantly gives incredible advantages not only in space! The techniques relating to this field allow doctors to monitor patients remotely, thanks partly to monitoring and sensing devices based on space technologies. Telemedicine has been fully integrated into every NASA human spaceflight program, including the ISS and the Artemis mission, and in order to do it the very best way, NASA has built strong partnerships with academic and private institutions.

In conclusion, NASA spin-offs create synergy between space exploration and medical progress. These advanced technologies are transforming medicine, improving diagnoses, treatments, and healthcare in general: exploiting their potential is a promising path towards the future.

R. G. Russo, *Navigating Health Challenges in Space*,
https://doi.org/10.1007/978-3-032-03620-9_30

Telemedicine is the provision of healthcare services where the traditional face-to-face patient-to-doctor interaction (or doctor-doctor) is replaced by long-distance interactions using information and communications technology (ICT), which stresses the role of unified communications and the integration of telecommunication, wireless signals, and computers. Telemedicine solutions are revolutionizing the healthcare industry through numerous applications, particularly in the fields of prevention, diagnosis, patient treatment, and monitoring.

The growing numbers of cases of chronic disease, heart failure, asthma, and cardiopulmonary disorders are accelerating the development and use of these advanced real-time patient monitoring solutions. They include teleradiology, teledermatology, tele-stroke/teleneurology, and telemonitoring for chronic diseases such as diabetes.

Improving public health and making the most efficient use of scarce resources are the main challenges for health infrastructures and logistics.

Telemedicine in Space [1]

It is now possible or the first time to delve deeper into cardiovascular physiology in microgravity and identify the main risk factors for astronauts during missions. This is now feasible using an ad hoc application for astronauts called GVM Assistance SpaceHealth, capable of monitoring numerous physiological parameters, with a wearable device. The bioparameters will then be analyzed during the week after return, in the recovery phase, and this will help define the changes in the astronauts' bodies during the low-gravity stay. One of the main challenges of space missions concerns the microgravity experienced in orbit and its consequences on the astronauts' bodies. With its special graphene and carbon fiber fabric, the wearable device causes variable compression on the astronauts' musculoskeletal system, promoting the correct distribution of body fluids and reducing the negative effects of microgravity, ensuring comfort, freedom of movement, and safety.

Space Safety Telemedicine

Space safety includes physical safety, communications safety, and space weather awareness. Physical safety includes avoiding launch and on-orbit collisions, minimization of human casualties from spacecraft or debris reentry, and the long-term sustainability of the space operations environment. The term *safe space*[1] refers to places intended to be free of bias, conflict, criticism, or potentially threatening actions, ideas, or conversations.

[1] A *safe space* in English is a place or situation in which you are protected from harm or danger; the term refers to places "intended to be free of bias, conflict, criticism, or potentially threatening actions, ideas, or conversations." (Wikipedia)

Space is often referred to as being very dangerous for humans, so what are the risks of venturing into this extreme environment? How can they be minimized? Is a space accident possible? And what influence does space travel have on a human being? [2]

Humans are not naturally equipped to live in space, so many precautions must be taken when leaving Earth. Technical and technological improvements have made it possible to avoid the inconvenience of certain old space equipment. However, the human body undergoes many changes: congestion, loss of taste, loss of muscle mass, and even weakening of the bones. Exposure to cosmic rays can cause changes ranging from cataracts to increased risk of cancer to infertility.

Much remains to be done to create a viable safe space environment for humans in the long term.

Man's numerous activities have meant that more and more objects already end up in orbit, increasing the risk of collisions and threatening operational satellites. Space weather also needs to be followed closely, as magnetic storms can damage sensitive electronic equipment and photovoltaic panels on satellites.

ESA develops satellites and sensors that monitor space weather better and help protect Earth from asteroids. Uninterrupted surveillance is essential for safe exploration and for the sustainable use of space.

A variety of hazards can exacerbate effects on the human body, which are being studied using ground-based analogs, laboratories, and the International Space Station (ISS). These locations all serve as test beds to evaluate human performance, as well as the effectiveness of strategies that could keep astronauts safe and healthy in space.

Through meticulous research, NASA is gaining valuable insight into how the human body and mind might respond during extended forays into space. The resulting data, technology, and methods form a knowledge bank from which scientists can extrapolate information for multi-year interplanetary missions.

The Future of Mankind in Space

Scientists are developing new tools for testing and manufacturing medicines and/or other medical remedies [3]. Yet our understanding of how the human body responds to medicine in space is still extremely limited. Scientists need to develop the pharmaceutical tools and medical understanding to support future missions in parallel. This will call for close collaboration from commercial space companies and space agencies across the planet.

Why Do We Go to Space?

Human space exploration helps us tackle fundamental questions about our place in the universe and the history of our solar system.

NASA's exploration provides some fundamental questions: Why are we here? How did it all begin? Are we alone? What comes next? At NASA, we explore the secrets of the universe for the benefit of all, creating new opportunities and inspiring the world through discovery.

An international partnership of five space agencies from 15 countries operates the International Space Station, and two dozen countries have signed the Artemis Accords, signaling their commitment to shared values for long-term human exploration and research on the Moon. Through space exploration, we gain a new perspective for our study of Earth and the solar system. New technologies improve our daily lives, and we inspire a new generation of artists, thinkers, tinkerers, engineers, and scientists.

Benefits to Humanity

Space exploration unites the world to inspire the next generation, make groundbreaking discoveries, and create new opportunities.

Shown below is a snapshot depicting some of NASA's ISS projects that could be of benefit to human health on Earth:

Additional space station research contributing to improved human health on the ground:

Project	Description	Benefits
ESA Brown-DTI	Tests the theory that the human brain issues previously untapped neural connections to adapt to space, examining changes in brain structure before and after spaceflights	Results show significant differences in upper-brain structures pre- and postflight that could affect long-term health and vision
ESA EDOS-2	Used a noninvasive technique tested in the EDOS investigation to examine whether astronauts continue to lose bone when they return to Earth and when the recovery occurs	Monitoring bone loss can inform diet and exercise programs used to counteract it in space and provide early detection of osteoporosis on Earth. The unique qualities of the scanner have led to its use in non-biomedical research applications
ESA ETD	Examined how microgravity affects the ability to keep the eyes on a fixed point while tilting or shaking the head. This ability relies on the vestibular system, or inner ear and nerves that maintain balance and orientation	The investigation used a special headset to track eye movement that proved ideal for tracking eye position in laser eye surgery without interfering with the surgeon, and is now being used in corrective laser surgeries through the world

Additional space station research contributing to improved human health on the ground:		
Project	Description	Benefits
IGAR	Developed by Canadian Space Agency and Centre for Surgical Innovation and Invention (CSii), this surgical robot acts autonomously after being programmed by a physician, performing highly accurate and minimally invasive procedures such as biopsy	A phase1 clinical trial showed IGAR is safe, accurate, and effective, and CSii is working on final US FDA approval. The technology could be applied to imaging and treatment for a variety of cancers, including prostate and liver, and eventually enable remote emergency surgery in isolated environments or during natural disasters
ESA MARES	A device that monitors muscle speed and force as astronauts exercise in an effort to counteract the decrease in muscle during spaceflight	The next generation of space exercise machines is likely to include variable counterforce capability, and MARES can be used for preliminary testing of the technique
ESA MUSCLE BIOPSY	Analyzes expression of nitric oxide, a signaling molecule linked to muscle activity, in human skeletal muscle before and after spaceflight	Improved understanding of loss of muscle mass in space could support development of more effective countermeasures for astronauts and provide insight into certain muscular conditions on Earth

Technologies and missions developed for human spaceflights have thousands of applications on Earth, boosting economies, creating new career paths, and advancing everyday technology all around us.

Coming Back to Earth

Whenever humans invest in space research, unexpected new technologies on Earth tend to follow. LASIK surgery, artificial limbs, cochlear implants, and infrared ear thermometers, among other healthcare innovations, all have their origins in NASA-funded research. The same could be said for drug development and manufacture, as the next era of research in aerospace medicine is sure to produce unforeseeable discoveries.

Benefits to Science

The pursuit of discovery drives NASA to develop missions that teach us about Earth, the solar system, and the universe around us.

Science at NASA answers questions as practical as hurricane formation, as enticing as the prospect of lunar resources, as surprising as behavior in weightlessness, and as profound as the origin of the universe.

Everyday Impact

NASA ensures inventions for space find practical uses closer to home. They often become answers to different challenges—the ones we did not set out to solve. These NASA "spin-offs" help detect breast cancer, protect firefighters, nourish newborns, and more.

Bibliography

1. Lewis RE. Risk of performance decrements and adverse health outcomes resulting from sleep loss, circadian desynchronization, and work overload, Mar 11, 2025.
2. Albornoz-Miranda M, Parrao D, Taverne M. Sleep disruption, use of sleep-promoting medication and circadian desynchronization in spaceflight crewmembers: evidence in low-Earth orbit and concerns for future deep-space exploration missions. Sleep Med. 2023;6:100080.
3. Loriè EP, Baatout S, Choukér A, et al. The future of personalized medicine in space: from observations to countermeasures. Front Bioeng Biotechnol. 2021;9:739747.
4. Krittanawong C, Singh NK, Scheuring RA, et al. Human health during space travel: state-of-the-art review. Cells. 2022;12(1):40.
5. National Academies of Sciences, Engineering, and Medicine. A new era in space: ensuring the future of biological and physical sciences research: a decadal survey for 2023–2032. Washington, DC: The National Academies Press; 2024.
6. NASA, ED. SGHEA G, WHY GO TO SPACE. Sept 22, 2023.
7. Catchpole JE. The International Space Station- building for the future. Praxis, UK; 2008.
8. Foss R. Food in the air and space: the surprising history of food and drink in the skies (food on the go), Paperback – May 25, 2016.
9. Doarn CR, Merrell RC. Telemedicine in space medicine and extreme terrestrial analogs. Mary Ann Liebert. 2014;20(5):405.
10. Nigossian A, Pober DF. Evolution of Telemedicine in the Space Program and Earth Applications, February 2001.
11. Martin A, Sullivan P, et al. Space medicine innovation and telehealth concept implementation for medical care during exploration-class mission. Acta Astronaut. 2012;81(1):30.
12. Urlings C, Baselet B, Tabury K, Baatout S. Virtual medical astronaut avatars for future deep space missions. Front Space Technol. 2024;5:1423138.
13. Krone B. Beyond earth: the future of humans in space, Paperback – May 1, 2006.

Part VI

The Future

31 The Future of Man in Space

Our Future Shapes, Our Present in Terms of Emotions, Fear, Future, Narratives, Opportunities

We have finally reached the end of this book. Evidently, the great conquest of space does not end here. Much time and energy will still be dedicated by man to what has always been configured as his most important adventure.

Space exploration has produced quite a few records of benefits for humanity. Space exploration has driven scientific and technological innovation that benefits people around the globe every day. Space exploration serves a cultural and inspirational purpose by fulfilling a deep need to understand the world, to address questions about the origins of life and the nature of the universe, and to expand the notion of what it means to be human. There is no activity on Earth that matches the unique challenges of space exploration. The first 50 years of space activity has generated benefits for people around the globe.

The actions we choose today will determine what our future will look like. What is not so obvious? Our future shapes what we do today. How can that be? The future has not yet happened, so how can it determine our actions today? How can this be true if we embrace the opposite view that our actions today shape our future?

We all have a personal view of the future, even though few of us make the effort to articulate that view, even to ourselves, much less to others. It is that view of the future that shapes our emotions today, and our emotions today shape our actions today.

Long space missions might have catastrophic outcomes for the health of astronauts. In addition to the already known risks of cancer, cardiovascular disease, and bone loss, a new study has revealed another deadly risk: the kidneys may suffer

Mircea AA, Pistritu DV, et al. Space travel: the radiation and microgravity effects on the cardiovascular system. Int J Mol Sci. 2024;25(21):11812.

R. G. Russo, *Navigating Health Challenges in Space*,
https://doi.org/10.1007/978-3-032-03620-9_31

irreversible damage. The combined effect of radiation and microgravity during spaceflight can lead to a variety of health problems for astronauts, including bone loss, muscle atrophy, and cardiovascular issues. Studies have shown that microgravity can exacerbate the effects of radiation on bone and other tissues, potentially increasing the risk of long-term health complications [1]. A prolonged period can have a catastrophic effect on the very delicate structures of all the body organs, especially on the tubules, which would be altered and damaged to the point of complete loss of their filtering function. Scientists believe that an astronaut returning to Earth may need dialysis to survive.

The researchers' aim is therefore to design drugs to protect the kidneys and other organs and systems from the risks inherent in prolonged missions in space beyond low Earth orbit. These might also be useful on Earth, for example, by increasing tolerance to chemotherapy drugs.

Is the human species destined to remain forever on its planet of origin or is it destined to expand into space? And, assuming that such expansion is possible and achieved, are there some insurmountable barriers beyond which we can never go?

Today, the indispensable instinct for discovery translates into a science fiction scenario, on which not only government bodies but also private companies have been working for years: space elevators that replace rockets at a tenth of the cost, experiments on the suspension and resuscitation of life for decade-long journeys into the cosmos, and solar sails that literally enable us to navigate in space. "The Future in Space" shows us how star travel will no longer be the exclusive prerogative of the military or NASA, but the logical destiny of generations to come because continuing exploration means guaranteeing our survival.

The instinct to explore and broaden one's sphere of knowledge, in the strictly geographical and intellectual sense, is a constant in man's character. If man decides to continue along the path he has taken so far, he will have to take account of the fact that expansion into space will most likely change him profoundly.

Bibliography

1. Shanmugarajan S, Zhang Y, et al. Combined effects of simulated microgravity and radiation exposure on osteoclast cell fusion. Int J Mol Sci. 2017;18(11):2443.

32 In Conclusion

Space medicine is providing us with new drugs and artificial intelligence programs capable of diagnosing diseases even before the symptoms appear; liquid biopsies that, with a single blood sample, can recognize the "signs" of different types of tumors; predictions of the evolution of diseases; and new telemedicine systems to intervene remotely.

All the innovations designed for astronauts will also be able to help us "terrestrials." This was discussed by the world's leading experts gathered in Florence, Italy, from September 13 to 15, 2025, at the Fortezza da Basso for the conference "Building a Space Faring Civilization - Advancing the Renaissance of Science, Medicine, and Human Performance in Civilian Spaceflight," organized by the Menarini International Foundation with the participation of NASA, SOVARIS Aerospace, and The Foundation for Gender-Specific Medicines. The event saw doctors, psychologists, and biologists face to face with astronauts, engineers, astronomers, historians, physicists, and ethicists to discuss in a multidisciplinary way the main challenges that await us in the new era of space exploration.

Our experts warn us that our existence on Earth faces an uncertain future: in fact, climate change, the depletion of important raw materials, collisions with other celestial bodies such as asteroids, and the invasion of new infectious agents are possible scenarios that make it imperative to explore alternative worlds on which to survive and prepare humanity for life on other celestial bodies. Preparing for this change is therefore fundamental from a medical, scientific, ethical, and political point of view.

The Main Threats of Space to Health

Space travel faces two main challenges: microgravity and oxidative stress, that is, an increase in free radicals to a level that compromises the cell's antioxidant capacity and causes DNA damage.

R. G. Russo, *Navigating Health Challenges in Space*,
https://doi.org/10.1007/978-3-032-03620-9_32

Research on astronauts has shown that oxidative stress derived, in particular, from ionizing radiation, which penetrates thousands of cells at high doses, alters the function of mitochondria and consequently the metabolism of carbohydrates and lipids. Furthermore, it damages DNA, modifies the expression of genes, and alters the length of telomeres. Furthermore, microgravity eliminates the impact of load on the bones and muscles, causes a loss of bone mass, and favors a redistribution of fluids toward the upper part of the body, increasing the risk of thrombosis and vision problems. As a result, the heart functions differently and loses contractility, while the left ventricle tends to become smaller and the walls of the arteries stiffen.

Space medicine is still in its infancy, but the good news is that we are light years ahead of precision terrestrial medicine as many know it. Space research is providing us with new tools to create personalized interventions on nutrition, physical activity, and medications in order to prevent disabilities. Besides, the analysis of biological samples on hair, saliva, breath condensate, and blood is also helping us understand the molecular basis of human physiology.

Ultimately

33

By ESA (European Space Agency)

The microgravity experienced on board the ISS imposes negative effects on the human body that can be compared to an early and accelerated aging process: if this may seem harmful, it actually represents a unique opportunity for medical research and science. To better support astronauts during current missions to the International Space Station and prepare for human missions beyond low Earth orbit, the Space Medicine Team is working to identify, evaluate, and develop new space technologies and procedures. The Space Medicine Team includes also the following domains: identification and evaluation of medical risks (including radiation) associated with low Earth orbit and deep space vehicles/habitats.

Shaping the Future

Finally, the experts at the conference discussed the numerous ethical issues that accompany the creation of a new space civilization.

The Mission

The development of technology—making new things to do new things—is central to ESA's existence. The Agency spends around 8% of its budget on direct research and development, an activity mandated in the ESA Convention. It runs many

R. G. Russo, *Navigating Health Challenges in Space*,
https://doi.org/10.1007/978-3-032-03620-9_33

preparatory programs to do this, covering various inventions, each leading activities based on their technical maturity or their field.

The Responsibility

To make sure that the right technology will be available at the right maturity at the right time for future missions.

MIX
Papier aus verantwortungsvollen Quellen
Paper from responsible sources
FSC® C105338

If you have any concerns about our products,
you can contact us on
ProductSafety@springernature.com

In case Publisher is established outside the EU,
the EU authorized representative is:
Springer Nature Customer Service Center GmbH
Europaplatz 3, 69115 Heidelberg, Germany

Printed by Libri Plureos GmbH
in Hamburg, Germany